W9-BVJ-277

STUDIES
IN HISTORY OF
BIOLOGY

# STUDIES IN HISTORY OF BIOLOGY

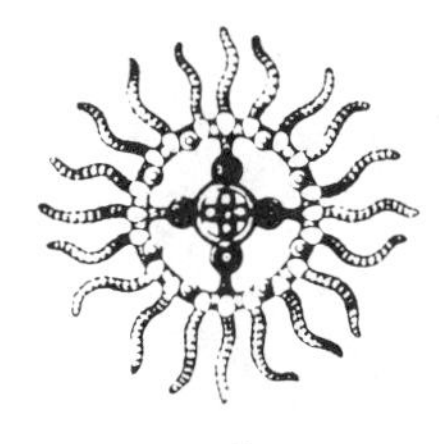

6

*William Coleman*
*and Camille Limoges*
*Editors*

The Johns Hopkins University Press
BALTIMORE AND LONDON

*Printed in the United States of America*
*The Johns Hopkins University Press, Baltimore, Maryland 21218*
*The Johns Hopkins Press Ltd., London*
*Library of Congress Catalog Card Number 78-647138*
*ISBN 0-8018-2856-2*

# Contents

STUDIES
IN HISTORY OF
BIOLOGY

# Darwin and the Laws of the Animate Part of the Terrestrial System (1835–1837): On the Lyellian Origins of His Zoonomical Explanatory Program

*M.J.S. Hodge**

## 1. Introduction: Explanatory Traditions and Programs and the Early Darwin

The immediate aim of this essay is a pretty narrow one.[1] From July 1837 on, Charles Darwin was theorizing, especially about the origin and extinction of species in relation to the "laws of life," in his Notebook B. Toward the end of that year, and that notebook, he spelled out very precisely the prospects for further inquiry that he saw opened up by the conclusions he had reached so far. This essay reconstructs how Darwin arrived at that view of these programmatic prospects, by analyzing successive developments in his theorizing from the time, nearly three years before, in February 1835, when he made his first break with his mentor Charles Lyell over the exchange of new species for old.

Narrowly restricted though the analysis is, however, it requires a preliminary clarification of certain broader issues if it is to be fully understood. For it deploys two general notions—of an explanatory tradition and of an explanatory program regarding organic diversity—that are designed to apply in other contexts as well as the Darwinian one.

The connection between the broader issues and the narrower analysis is direct. Everyone, or nearly everyone, is now against "Whiggish" interpretations of history, especially those that search through earlier authors for "precursors" of later ones. What is proposed, here, is that a general

*Division of History and Philosophy of Science, Philosophy Department, Leeds University. 

antidote to any such interpretations can be applied to the particular case in hand; so that we can have a definite and deliberate alternative to approaching the young Darwin, of the notebooks, as a precursor of the older Darwin of the *Origin of Species.* We should consider next, then, how this antidote is designed to make that alternative possible.

There will be little disagreement that Aquinas's *Summa Theologiae* (Pt. 1) and Buffon's *Epoques de la nature* are in a serious sense addressed to the same subject. For not only do they both talk of "species," they both propose or presuppose answers to such questions as why there are many kinds of animals and plants rather than none or a few; how these kinds differ from one another; how the diversity among species may be referable to certain sources that are somehow more original than the species themselves; how there can be many individuals of each species; how the first members of each species may have come into existence; and so on.

This common concern with such questions is easy enough to concede. But there are several historiographical responses that we might make to it. We need to distinguish two.

We might respond as did those defenders of Darwin (Huxley, Haeckel, and Clodd, for instance) who rewrote the history of Western thought on behalf of their campaign. We would then fit Aquinas, Buffon, and any other pre-Darwinian authors into a comprehensive see-saw scenario wherein "evolution" is seen to "rise" as some contrary to it—Hebrew "creation" or Hellenic "stasis"—falls.

Now, the trouble, of course, with relating all earlier authors to some single trend, such as the "rise of evolution" and decline of its contraries is supposed to be, is that any issues about those authors' own intentions are either ignored or begged. For, in the very decision to ask of everyone—Empedocles, Aristotle, Ray, Maupertuis, or whoever—whether he was ultimately a forerunner of Darwin, or rather a forerunner of Darwin's opponents, is at best to ignore his own understanding of what it was he was trying to do; at worst, to imply that he was trying, unsuccessfully, to write the *Origin*, or trying, equally unsuccessfully, to ensure that it never be written.

So, the drawback to precursor historiography, here, is what it always is: it avoids or distorts the interpretation of intentions; not surprisingly, since it is impossible, in principle, to intend to be a precursor. No less obviously, then, the same flaw is inherent in any sorting or mapping, whether of individuals or institutions, in relation to any enterprise or outlook—"evolutionary biology" or "an evolutionary view of nature"—that is held to come of age only later, in the work of Darwin. Such sortings and mappings inevitably suppress interest in what those authors understood themselves to be contributing to, in favor of interest in what they did or did not contribute to the enterprise or outlook whose career the historian has chosen to chart.

That the drawback is a quite general one can be seen by noticing its analogues elsewhere. A century ago "thermodynamics" was another new venture that was supplying itself with a suitable history. But, of course, to seek an account of "thermodynamics" in the Renaissance or the Enlightenment is to avoid the very issue of how the subject became constituted as it was, in the 1850s, when laws of thermodynamics were first formulated as such.

So, going back for a moment to the *Summa* of Aquinas and the *Epoques* of Buffon, could there be a historiographical response to their common concern with questions about organic diversity that would neither ignore their authors' intentions nor interpret them only in reference to some one future development? Clearly there could be, provided only that we were prepared to relate them, in each case, to one or more among many earlier developments that were identified in an appropriate way. And this we could do by relating them to one or more among the many explanatory traditions that there have been in the understanding of the problems taken to be posed by organic diversity. For these traditions could be identified by the problems they took to be most fundamental, and by the resources they took to be most indispensable in solving them.

Thus, for Aquinas's understanding of the problems posed by organic diversity, the Platonic is the decisive tradition. For he takes as his fundamental problem explaining why a single, simple God could and would create many, differing creatures; and he seeks his solution in the argument that God can do so, because the singularity and simplicity of his essence is consistent with that essence being imitable in creatures in many diverse ways existing as so many distinct Ideas in the divine mind; and in the argument that God would do so, because he creates so that his goodness shall be communicated to creatures and represented in them, and because that goodness can only be adequately represented by a multitude of creatures distributed in a scale of perfections, from the lowliest of the corporeal to the highest of the angelic species.

Now, in these familiar Platonic moves of Aquinas, we see a feature likely to be distinctive of any explanatory tradition regarding organic diversity: a concern with a characteristic set of primary reconciliatory problems. For the diversity of animal and plant species will have to be explained by showing how it could derive from a source or sources, a principle or principles, an origin or origins, less diverse in some respects than the species themselves. An account must be given, therefore, as to how the simplicity, singularity, homogeneity, or uniformity of the one is consistent with the plurality and variety of the other.

So, too, with Buffon, who is (the silence of his historians notwithstanding) explicitly concerned with two such reconciliation problems. For he has to square the structural and functional diversity among species with the qualitative identity of the ubiquitous "organic molecules," their lack of any intrinsic differences being required by his theory of generation. And here his principal explanatory resource is his account of how the brute matter associated with those organic molecules can differ in its blend of the four elements; so that the resultant associations can exert on nutrient materials different patterns of active, organizing forces, and so can constitute diverse, more or less stable, self-reproductive "internal molds." Moreover, beyond that, Buffon—as Newtonian as Aquinas was Platonist—has the still more fundamental problem of reconciling the very distinction between brute and organic matter with the inertial and gravitational homogeneity of all matter taught by the *Principia* of Newton, as interpreted by writers like Cotes, a problem Buffon resolves by invoking the expansive effects workable by a special matter of fire, which, though distinctive in that power, shares too in inertia and gravity and ultimately owes that expansive power to those properties.

Turning from traditions to programs, we see, then, that while someone's primary reconciliatory problems may derive from a commitment to one or more longstanding explanatory traditions, his program, his agenda, may well be developed as an unprecedented attempt to tackle a new situation encountered in quite novel circumstances. Aquinas was knowingly upholding the Platonic tradition even while he attempted what no one, even his own teacher Albert, had previously undertaken: a full reinterpretation of the Genesis narrative as the original creation ex nihilo in six (literal) days of an Aristotelian cosmos.

In confronting the new challenges that this program entailed, Aquinas's problems were often integrative ones: problems in integrating Aristotle's teachings with biblical texts, conciliar edicts, and the Augustinian account of the divine Ideas. And such integrative problems are, indeed, typical of such programs. Buffon's *Epoques* sought—through a reconstruction of the earth's declining thermal and constant gravitational economy—to integrate for the first time the microcosmogony (his theory of animal and plant generation) and macrocosmogony (his theory of the earth's formation as a habitable terraqueous globe) that he had developed separately some thirty years before.

The bearing of this general discussion on the case in hand, Darwin in 1835–37, will now be apparent. For the aim here is to bring out how Lyell's proposals for reforming geological science as a whole included articulating a new tradition in the understanding of the problems posed by organic diversity; to clarify the explanatory program Lyell himself pursued within this tradition; and to show how Darwin continued this same tradition, even while a succession of disagreements with Lyell led him eventually to develop a novel explanatory program of his own.

The analysis will, therefore, contrast with two older tendencies in Darwin studies (both prevalent in the centennial days around 1959) while reinforcing another, more recent trend.

According to one older tendency, the *Origin of Species* was read as "evolution plus natural selection as a mechanism for it," and then a search was made in the notebooks for the appropriate precursor, by asking when and why Darwin first "believed in evolution" and when and how he first "discovered natural selection." According to another older tendency, Lyell's *Principles of Geology* was read as "uniformitarianism in geology," and Darwin's *Origin of Species* as "evolutionism in biology"; and then it was inquired whether Huxley was right in holding that the first demanded the second, or whether, on the contrary, it precluded it, with the corollary that Darwin's "evolutionism" may have come from the "progressionist" opponents of Lyell's "uniformitarianism."[2]

Both these tendencies have had to be dropped of late. For the great increase in our knowledge of Darwin in the 1830s over what was current at the time of the 1959 centennial has shown that the assumptions and distinctions implicit in these tendencies are inadequate bases for understanding his thinking in the years of the *Beagle* voyage (1831–36) and the B–E notebooks (1837–39). The monographic studies of Herbert, of Gruber, and of Kohn have made especially clear why they should be inadequate.

Herbert explored the many interactions between Darwin's Lyellian theorizing about land elevation and subsidence and his successive conjectures about species origins and showed that any abstract, categorical comparison—or contrast—between "uniformitarianism" and "evolutionism" could only introduce an unhelpful oversimplification into what was itself a very complex matter.[3]

Gruber emphasized that the notebooks (from B on), exploring the means whereby new species originate in the transmutation of old ones, always embedded such thoughts in an explicit, comprehensive framework of argumentation—about the more ultimate causes, both "physical" and "final," of all the changes undergone by the animate world, from simple remote beginnings to the recent origin of conscious and conscientious man.[4]

Kohn has detailed how, even on the particular issue of the physical causes of species origins, Darwin's thinking went through several phases of elaborate speculations prior to his reaching the theory of natural selection, with later phases often repudiating what earlier ones had concluded.[5]

One implication from all three of these studies is, then, that we should ask not how Darwin was somehow drawn, in the years 1835 through 1838, toward transmutation and later natural selection. We should ask, rather, how he came to move, on many issues at many levels, away from his 1834 acceptance of Lyell's views in the *Principles*.

The present essay will disagree from time to time with Herbert, with Gruber, and with Kohn, just as they have disagreed with each other. But in this essential policy it will be confirming their implicit consensus; while it draws throughout on the editions, interpretations, and datings already given, by these and other early Darwin specialists, of the rich but difficult textual sources on which the newer understanding of Darwin's thinking in those years has been grounded ever more firmly in our time.[6]

## 2. Lyell, the Theory of the Earth, and the Exchange of Species

Turning, now, to Lyell, there will be no disputing that it was his aim as a geological theorist to develop a neo-Huttonian system of the earth and not merely to square it, but even to support it, with the post-Huttonian discovery of faunal and floral succession. Nor will there be any disagreement that this was a quite new aim; to Lyell's own contemporaries, indeed, what was surprising was not that it had not been attempted before, but that anyone, even Lyell, was attempting it then. For it was widely assumed that the paleontologists' recent discovery of faunal and floral succession had finally made the prospects for any version of the Huttonian theory quite hopeless.[7]

Hutton's system of the earth had been designed by Hutton to explain (and by Hutton's God to ensure) that aqueous and igneous causes of change acted and interacted with undiminished intensity in the long run, so as to keep the earth's surface constantly fit, indefinitely far back into the past and forward into the future, for every single species of organism that is living upon it now. But between Hutton's finishing his *Theory of the Earth* (1795) and Lyell's beginning his *Principles of Geology* (the three volumes appeared respectively in 1830, 1832, and 1833) in the 1820s, almost all geologists had come to accept, not only that many species had become extinct over vast eons of the past, but that they had been replaced by new species at various times. Several distinct faunas and floras had thus succeeded one another; there had been several complete turnings over, several revolutions, of organic life.[8]

Hence, then, Lyell's novel reconciliation problem. In developing his own stable, balanced neo-Huttonian system of aqueous and igneous causes destructive and productive of continental lands and oceanic seas, Lyell had to explain how these physical causes indefinitely ensured the general habitability of the earth—its continued habitability for *some* species representing the two dozen or so main types of life (mosses, mammals, conifers, mollusks, and so on)—and yet how those same causes ensured, also, that its habitability for any *particular* species should not be conserved. He had, that is, to reconcile the uniformity, the permanence, the conservation of the earth's general habitability with continual innovations in the species of its inhabitants, with a continual exchange of new species for old; for each species really is new; it has never lived on the earth before and once it is extinct it will never return.

Naturally, we learn most about Lyell's understanding of this problem by following the many arguments he develops in the course of trying to solve it. But those arguments are sufficiently elaborate, not to say tortuous, that we will do well to consider first their common premises and conclusions, before we pursue them through their many inferential twists and turns.

We do best to start with what is causally prior to all the rest—the system of aqueous and igneous agencies of change. Here, as elsewhere, we need to appreciate Lyell's proposals as to both the total effects wrought upon the planet as a whole and the changes brought in the short and long run to any particular region. Aqueous causes level the earth's surface, while igneous causes counteract this leveling by producing at some times and places subsidence and at other times and places elevation of the land above or below rivers, lakes, and seas. For the globe as a whole in the long run, subsidence and elevation are in balance with one another, and in their joint unleveling effects they are balanced by the leveling power of water. So the total amount of land and of sea may fluctuate widely but never increases or decreases irreversibly. Throughout these fluctuations, subsidence and elevation as agencies are not permanently restricted to any special sites. Rather they rove very slowly about all over the globe and will eventually visit any spot again. The distribution of land and sea around the globe is therefore constantly changing. A region that was once occupied by a continuous continental mass of land will later be open sea, a chain of freshwater lakes, or an ocean archipelago and eventually continental land again. Following as they do from successive visits of the basic agencies of subsidence, elevation, and leveling, these changes are themselves slowly roving about the globe, circulating, not of course on a fixed track, but like a hostess circulating at a party, moving from one site to another and so bringing to the whole arena a characteristic repertoire, a characteristic cycle of change.[9]

These changes in land and sea entail changes in the planet's overall mean temperature, because the larger the proportion of the total land close to the equator, the larger the proportion of the incident solar heat absorbed, given that land absorbs more heat than ocean does; likewise, the more land masses are clustered around poles, the lower the earth's mean temperature. Such changes can be large, but dependent as they are ultimately on the stable system of igneous and aqueous causes, they will not continue irreversibly in one direction.

Since it is derived from the cycle of land and sea changes, the theory of climate shifts is cyclic too, in that, as Lyell says himself, any mean temperature for the earth will eventually return with the return of an appropriate distribution of land and sea in relation to the poles and the equator. Lyell, indeed, emphasizes this reversibility by calling the extremes of mean temperature, caused by clustering of land around the equator or poles respectively, the summer and the winter of "a great year." However, although there is reversibility, there is no periodicity. For there is no recurrent succession of climates, much less any periodic sequence. Any autumn may turn out to be as well called a spring in that it is as likely to be followed by a shift back toward a summer as by a shift on toward a winter. Thus in Lyell's reconstruction of the climatic succession, from the time of the earliest known fossiliferous rocks (laid down in the Carboniferous age) to the present, there is a shift away from one extreme to another. For the mean temperature of the earth has made in this whole time a continual, usually very slow but sometimes briefly quickened, progress away from a summer and toward a winter. Naturally, the theory allows, indeed, requires in the indefinitely long run of the future for this shift to be reversed; but it does not require any such reversals within the limited, recoverable history recorded in the rocks. Europe was tropical in the coal age, then, Lyell suggests, because there was a clustering of large land masses around the equator, while the land where the European continent now is consisted largely of islands.[10]

The leading thesis of the *Principles of Geology*—the adequacy of causes presently active for all changes recorded in the rocks—is thus saved from one obvious objection. The higher temperatures in the northern regions of old need not be explained as a residue from a primitive, molten fluid state of the globe, a state that causes active now could never reproduce.

That thesis is also saved from another obvious objection, Lyell argues, once one supplies the further premise that the representation of supraspecific groups of animals and plants is determined over time, as it is over space, by adaptation to climate. Islands in tropical oceans today have many reptile species but few if any indigenous mammal species. The absence of mammal remains in the European Carboniferous strata may, then, be a record of a condition and stocking of the earth that existing causes could reproduce with enough time in the future. Conversely, the absence there of such remains need not be taken as a record of their total absence from the whole planet at that time, nor therefore of a total absence supposedly due to the whole planet then being unfit for mammals, thanks to a high mean degree of heat supposedly only explicable as a residue from its original formation as a molten fluid globe.[11]

With these conclusions—as defended by Lyell in his first volume—we are prepared to appreciate what is most distinctively novel about his conclusions as to the changes now going on in the animate world, as he presents those conclusions in his second volume: namely, his combined concern with questions about species and with the adequacy of current causes for the past effects recorded in the rocks. It may seem so obvious that Lyell should combine these two concerns, that we would expect others to do so too. But a moment's reflection, and a glance at how Sedgwick, for instance, was then approaching such subjects, should save us from that mistake. Sedgwick, too, especially in his presidential addresses to the Geological Society in 1830 and in 1831, was concerned with the question of whether past faunal and floral succession had taken place through sudden, wholesale extinction and replacements of entire faunas and floras, or rather through gradual, piecemeal exchanges achieved by one-by-one extinctions and origins of species. But, he deliberately declined to seek one, single general answer to this question, and insisted, rather, that in some formations, there appeared to be a manifest record of sudden, wholesale exchanges, while in others there were unmistakable traces of gradual, piecemeal ones.[12] There is, thus, in Sedgwick's approach to the issue, no place at all for the very inquiry Lyell makes prior to any attempts to reconstruct past changes from their memorials in the rocks: an inquiry into whether there are now causes entailing the exchange of new species for old, causes adequate, given enough time, to produce the sorts and sizes of faunal and floral change recorded in the rocks.

Lyell's fundamental commitment to the adequacy of existing causes brought him, therefore, not merely to a preoccupation with current changes, whether physical or organic, that Sedgwick could not share, but to a peculiar preoccupation with species, which no one pursuing Sedgwick's alternative to Lyellian geology would see as legitimate.

Lyell pursued that enquiry through eleven chapters of his second volume and on into his third. And he concluded that there is a continual, ubiquitous exchange of species going on, though not at a precisely constant rate because of its complex dependence on physical change. The conclusion is not, however, that this exchange is directly observed in action at present. It is rather that nothing observed in the three millennia of the historic period is inconsistent with the supposition that it is going on, though too slowly to have come before the observing eye; and that this exchange is the only supposition reconcilable with the stratigraphical paleontology of the youngest formations, the Tertiary ones. So, on the general presumption that all past causation is continuing tirelessly into the future, it is presumably going on now.[13]

No one before Lyell had had his reasons, nor indeed any reason, to inquire at that length, or indeed at any length, whether such a conclusion might be evidenceable in that way. Equally, then, no one had had his reasons for tackling as he did the four clusters of questions about species that he took up first in making his case for it, the four clusters of questions being: (a) whether species are fixed or mutable; (b) whether, if fixed, each traces to one small original stock and so to a single origin in space and time; (c) whether, once it has originated, the duration of each is limited by its dependence on changing conditions; and (d) whether currently observable physical changes entail that the many species extant at any time eventually become extinct successively, that is, one after another rather than in batches together, and, if so, whether these successive losses of species are compensated by a successive origin of new species likewise now in progress.[14]

Lyell's answers to these questions are, of course, familiar enough and in any case often implicit in those conclusions of his that we have already recalled. Species are fixed, and each does trace to a single original stock; their durations are limited by conditions, and their successive extinctions are compensated for by successive origins. No less familiar is an asymmetry in Lyell's exposition. He finds causes observably active at present adequate to effect successive extinctions in the long run. But the causes of species origins he leaves quite mysterious not to say miraculous. On his account, we do not witness, or at least have not witnessed, directly either species extinctions or species origins; but while he commits himself on what we would see if we were to watch an extinction, he makes no such commitment concerning species originations.

What are important, here, however, are not only the conclusions themselves, but the resources Lyell introduces, the inquiries he undertakes, and the status he gives the conclusions in relating them to one another and to the overall aims of his reform for geology.

His treatment of his first question concludes that although there is some capacity in all species to acquire adaptive and inherited characters in changed conditions, and although this capacity is much greater in some species than others, it is always limited so as never to allow the transmutation of one species into another.

With no continuous transitions between species over time, specific distinctions among individuals are real not arbitrary, and so he can go on to ask whether all the individuals of a species descend from a small original stock, perhaps even a minimal first pair, and consequently originate at a single location. Since the single stock hypothesis implies this geographical law as to "the first introduction of species," it must be reconciled with known generalizations about the actual distribution and powers of migration of the species extant now. In reconciling it with these generalizations about distribution, Lyell holds, one must start by following Linnaeus in distinguishing those generalizations relating to the *stations* (roughly equivalent to the modern terms "habitats") and those relating to the *habitations* (roughly like the modern "ranges") of species. For the station (e.g., temperate salt marsh) of a species is the complex of conditions—the climate, the soil, the food, and so on—that must be present in an area if the species is to live there; the habitation (e.g., Northern Europe) is the area it actually occupies. So, Lyell stresses, following De Candolle, what determines a species' station, determines what it needs to live, is the fixed physiology constituting its specific character; while its habitation, the area it occupies, depends on where it originated and what areas have been not only suitable but accessible to it since. Its habitation depends, then, on the changes in land and sea, in climate and soil, and in other animal and plant life, recorded in the rocks and studied by geologists.[15]

Lyell is prepared to argue that the hypothesis of single stocks is consistent with what we observe today; for it can explain all the known generalizations about geographical distribution, if suitably supplemented with independently evidenced generalizations about the constant physiological determination of stations and the shifting geological determinants of habitations. It is this argument that leads to the extinction question being subsumed within the treatment of biogeography. For biogeography, as an enquiry into the causes limiting the spatial spreading of species at present, must indicate what causes limit their temporal duration by limiting how long they can hold their ground in a world subject to the ceaseless agency of igneous and aqueous causes. Lyell's account of extinction presupposes, therefore, that the explanatory resources provided by De Candolle's biogeography are adequate, when integrated with those provided by his own neo-Huttonian geology.

Lyell's treatment of extinction is thus a prime instance of what we may call his first actualistic strategy. The motivation for Lyell's actualism, his preference for presently active causes, is his insistence on the direct observability of explanatory causes. Only those active in the human period, and so the period of human observation, are directly observable, in principle. Hence, then, the rationale for the uniformitarian premise: that while the present is special in the observability of its causes, it is not special in any other way; so that those causes and no others may be presumed active at earlier, prehuman times. Lyell can, then, meet the demands of his actualism by finding for a given explanatory challenge, recorded in the rocks, such as species extinctions, causes observably active today whose short-run effects in the present show them to entail such consequences in the long run of the future.

However, failing that, there is another way to meet those demands; and we may call it his second strategy, as he explicitly views it as a second best to be resorted to when the first cannot be followed.[16] According to this second strategy, one finds in the rocks traces of the past effects of some changes, the successive origins of new species, say; and one then shows that the presumptive continuation of such changes through the present and on into the future is consistent with what we observe today. Thus, although not observedly in action today, they could be going on now and so be observable in principle, however unobservable in practice. Thus, the first strategy sticks with changes that are already known, observedly, to be going on now; the second retreats to those that are not but could be, because they may be presumed to be going on now observably but not observedly, as they were unobservably in the prehuman past, when they were producing the effects recorded in the rocks.

We can appreciate the importance of these two strategies for questions about species replacements when we also appreciate the importance for Lyell of an analogy between his treatment of species replacements and his treatment of continental land replacement. His account of continental land replacement was decisive for his entire theory of the earth, as that theory was developed by him in removing what he saw as a flaw, a lapse from the actualistic perfection, in the Huttonian system in its original version. For, Lyell held (probably wrongly) that there was a glaring discrepancy in the older Huttonian version (he was thinking of Playfair in particular). The destruction of past continents was credited to causes working continuously, and so, observably, even observedly, through the present and on into the future. But the replacement of these old continents, with new land masses formed in the beds of the sea, was credited to occasional, widespread, paroxysmal upheavals in the past, which—although ascribed to tireless igneous agency, and so expected to occur in the future—were supposed not to have come within human experience of the present. Lyell accordingly labored hard over earthquake and volcano records from the historic period (the present) to avoid any such discrepancy in his own system.[17]

Likewise, then, with the possible replacement of species, an analogous issue that this latter-day neo-Huttonian must now face. If one takes it that the successive extinction of species is "part of the constant and regular course of nature," then the question obviously arises, says Lyell "whether there are any means provided for the repair of these losses?" Does the "economy of our system" include an endless depopulation of land and sea; or does it, rather, include periods of depopulation like the present, followed eventually by a new era "when a new and extraordinary effort of creative energy is displayed?" Or, again, do new species now originate from time to time but without naturalists catching them in the act?[18]

The problem is pressing and legitimate, even though Humboldt has said, as Lyell reports, that such subjects are mysteries which science cannot reach, and that zoological and botanical geography does not "investigate the origin of beings [i.e., of species]." To geology, Lyell counters, meaning of course, his actualistic, uniformitarian science of geology, "these topics do strictly appertain." For, he says, this science looks to the geography of the present only in order to relate it to the different geographical situations of the past. Hence, as Lyell implies, the geologist must at least ask whether the origin of species, a kind of event known to have happened in the past, is still occurring at present. However, he goes on to show, even if there were an extinction and an origin of a species somewhere in the world as often as once a year, it would be only every eight thousand years or more than a land mammal species would end or begin its life in Europe; so even such a high global rate of species exchange is consistent with no conspicuous one having come or gone in historic times in the countries closely observed by naturalists. Perhaps naturalists' observations in future centuries may eventually provide the data adequate for inferring directly "the laws which govern this part of our terrestrial system," the exchange of old species for new, that is. Meanwhile it is from the geologists' findings in the recent strata, including fossil remains of living species, that these laws must be inferred indirectly, in a second strategy inquiry.[19]

When these strata are ordered chronologically, on mineralogical and other criteria independent of their fossil contents, we can learn which extant species have survived most physical changes and most extinctions of other species, and which originated more recently, when the earth was much as it is now. From these facts we can decide, Lyell explains, whether the hypothesis already shown to be consistent with existing data and with the lack of data from historical records, the hypothesis that there are always and everywhere origins of new species compensating for extinctions of old ones, can be confirmed as being, indeed, a law of the uniform system of terrestrial nature; for from such data we may be able to conclude "whether species have been called into existence in succession" rather than all at one period; "whether singly," rather than in "groups simultaneously," and also whether "the antiquity of man" is such as to make our species one of the youngest, rather than one of the oldest now extant. Thus will Lyell later conclude, in his third volume, that when these questions are settled by examination of the Tertiary formations, there is little doubt that this hypothesis does identify one of the laws that have governed and so, presumably, do still govern the exchange of old species for new.[20]

In due course, we shall be coming back, and more than once, to Lyell's treatment of these subjects. But we have seen enough to appreciate that a zealous disciple of Lyell's, such as Darwin saw himself to be, in 1835—one, moreover, with an intense interest in continental land elevation and in the extinction of species—would be engaging a combination of problems and resources for their solution without precedent in any other explanatory tradition regarding organic diversity. So, we may move now from considering the *Principles* as a text and a publication of Lyell's own, to consider the context it was providing for Darwin's private reflections on his geological findings, including the biogeographical ones, in South America.

## 3. Darwin as a Lyellian Geologist in 1834

For a start we need to clarify the relation between Darwin's geological notes and other documents dating from Darwin's *Beagle* years (late 1831 to late 1836); Darwin's personal development in the first three of those years; and his debts to Lyell, not only in general theory but also in the specific interpretations he was giving of South American geology and biogeography.

Unfortunately, for any attempts to keep anachronism to a minimum in our understanding of Darwin's thinking on his voyage, the documents from these years are not as familiar or as easily studied as others, like the *Journal of Researches* (1839) and *Geological Observations on South America* (1846), where Darwin gives us his later theoretical reflections on his earlier findings. This makes it imperative not to forget one obvious truism about the Beagle years, namely, that Darwin could well be excited for theoretical reasons by some finding, when first making it, a fossil or a peculiarity in some animal's distribution, say, without these reasonings having much in common at all with those he will later base upon it. The importance of this point is all the more obvious when we recall the different kinds of documentation that Darwin was preparing during his voyage. Apart from a diary appropriate for reading by the family on his return, his regular writing duties included (a) field notebook entries; (b) animal, plant, and rock specimen lists; and (c) two scientific diaries, one of geology, the other of zoology.[21] Darwin often made later, telling insertions in all of these, and he sometimes cut out pages from the notebooks or separated some of the loose quarto sheets of the geological and zoological diaries when drawing up drafts of essays or chapters for books for possible further revision on his return to England. But despite these sources of uncertainty, a fairly clear impression emerges of his general development as one reads these documents from the voyage years together with the letters home to friends, to family, and to fellow men of science, Henslow especially.

A conveniently memorable coincidence for us here is that in May and June 1834 Darwin sailed around the southern tip of South America and so left the eastern plains for the western mountains once and for all, just halfway through 1834, and halfway through the five years he was abroad. By this midpoint, he was a very different person from the young green graduate who had been taken on originally by Fitzroy. In matters geological and zoological he was already thinking of himself as a budding expert, ambitious for professional recognition; in matters social and political he had found a vast gap between his own ideology and Fitzroy's conservative, authoritarian outlook; in matters religious he had lost his earlier biblical literalism but kept his faith in the natural theology of providence and the revealed theology of salvation; and in matters of civilization and humanity he had had to face the fact that all mankind were not even Europeans, let alone British officers and gentlemen like himself and those he messed with on board.

A comprehensive view of Darwin at this time would have to look for possible links between these various developments in his character and experience. Although a general study of such links is not needed here, we can see that even in considering Darwin's zealous commitment to the principles of Lyell's geology, we would have to take several if not all of them into account.

Few things concentrate the mind of a student more than a fundamental challenge to the views of his first teacher. Before Darwin boarded the *Beagle* complete with a copy of Lyell's first volume, his teacher in geology, Adam Sedgwick, had explained, in his February, 1831 *Address as President* to the Geological Society, how Lyell's views in that volume were radically at odds with his own. Moreover, Sedgwick championed, as a theory directly contrary to Lyell's principles, De Beaumont's proposal that the mountain ranges of the world had been elevated suddenly at a few special periods in the past, by causes far more powerful than observation allows us to think in action still today. When Darwin read Lyell's third volume, after returning from his Santa Cruz River expedition in April 1834, he found De Beaumont's theory expounded and attacked at length by Lyell, as one rightly seen by Sedgwick to be quite inconsistent with the principles of the *Principles of Geology*. He also found a specific part of De Beaumont's general theory singled out for separate criticism: the conjecture that the Andes mountains were the last range to be so raised, and that they were formed in a single vast upheaval that caused a great oceanic wave and a rising in the oceanic water level identifiable with the deluge recorded in many historical traditions around the world. There was, then, really no way that Darwin could have geologized all over South America without sooner or later taking his stand either with Lyell or with De Beaumont. So, he was eventually going to have to make a decision on the fundamental matters of principle dividing Lyell and Sedgwick.[22]

There is abundant evidence from the field notebooks, the geological diary, and the correspondence that Darwin's interpretation of South America was still largely consistent with Sedgwick's and De Beaumont's ideas late in 1832 but that well before he rounded the Horn, in June 1834, he was following Lyell. For his 1834 reconstruction of the history of the eastern plains required him to reject anything like De Beaumont's account of the western mountains. More precisely, in 1832, when explaining the plains and their fossils, he had supposed that a few vast elevations and resultant diluvial waves from the west had deposited in the sea the gravel and other rocks later elevated to form the plains, and he had supposed another flood to have extinguished many species of extinct mammal quadrupeds and to have swept their bones to the sites where they are now found as fossils. As we have seen, sudden floodings of lowland plains were an allowable part of any Lyellian's explanatory repertoire, because observable causes currently active (the retreat of the gorge at Niagara was Lyell's favorite example) would certainly produce them in the future. But to make the deposition and later elevation and flooding of lowland plains dependent on the sudden formation of a high mountain range in a quick series of vast upheavals would be unacceptable, because nothing we observe today warrants our predicting such sudden, massive elevations as the consequences of the igneous causes presently at work.

Darwin's rejection of his 1832 views and his adoption instead, by 1834, of a great many little upheavals spread over a long time, and raising both the plains and the Andes one small step at a time, marks, then, the decisive stage in his adoption of Lyell's teachings in physical geology. For the same sources show that his conversion was later completed by his knowingly following Lyell on another no less decisive issue. Observations made in the Santa Cruz valley in April 1834, when interpreted immediately afterwards with the help of the Huttonian account of successive primary rock production by metamorphosis, elaborated by Lyell at the end of his third volume, provided Darwin with a final confirmation of his many small elevations for both the plains and the mountains.[23]

Naturally, he had reflected constantly on the implications of his new views of the physical geology of the region for the interpretation of its extinct and extant inhabitants. But he had reached no settled conclusions in 1834, except the one drawn from comparing the mollusks living in the sea with the fossil mollusk shells scattered on the coastal plains. For these fossils still included some organic material and were of the very same species as those living at the coast; so they confirmed Lyell's insistence that the species now extant had many of them lived through important changes in physical geography and that the agencies which had earlier produced the earth's present land and sea had not once acted with a violence fatal to all the species living in the areas they were then remodeling.[24]

What Darwin had found it impossible to settle, even by 1834, despite repeated wrestling with the data and various interpretations, was the relative chronology of the extinct mammal and the extant mollusk species. The question was a complex one because it required a decision as to where the mammals had lived, whether on the coastal plains near where they were now found as fossils, or whether to the west in foothills or mountains whence their remains had been carried as whole carcasses or separate bones, and whether before or after the first elevations to raise the ocean bed into low-lying land. In January 1834 at Port St. Julian on the Patagonian coast, he did in fact make a fossil find that he was eventually to interpret as settling these questions. But over a year was to pass before, in February 1835, and probably when sailing between Chiloe and Valdivia on the west coast, he wrote out the relevant reasoning in full. It was this reasoning to resolve these difficulties in his Lyellian interpretation of the region that brought him to face directly general difficulties in Lyell's account of extinction.

## 4. Darwin's Problems with Extinction (1835)

Lyell's account of the causes of extinction follows his first actualistic strategy. He begins with what is going on at any place from day to day, from season to season, and then moves outward in space and onward in time to show what will happen over the whole globe from one epoch to the next. At any spot only a few of the species are present that could flourish there, while some at least of the others are pressing in upon them, having been kept out precisely because those present have so far occupied that ground and have so far had some slight advantages over the encroaching neighbors in the complex struggle for their common vital requirements. But the competitive balance in this struggle is highly sensitive to many disturbing factors. Some of these are physical factors—soil, temperature, altitude, and so on—others, often even more decisive, arise from the presence of yet other plants and animals in the area that are supplying the competing species with food, hosts, shelter, parasites, and so on. These competitive balances are therefore directly or indirectly disturbed when any other species expands its range into the area. In this disturbance some species will likely lose ground and numbers; indeed, certainly, in a region already stocked to capacity; for there the new arrival of an invading species, no less than the permanent increase of one already present, must entail a loss of population for some other species.

By studying the successive reports of travelers and naturalists, especially in the Americas, over the centuries, Lyell shows that such local, partial exterminations have often followed the disturbances in the balance of nature caused by the spread of human populations and the attendant domestic quadruped species. But similar interactions between species that must eventually lead to similar effects are observed, too, in areas where man is barely involved at all. The conclusion must be, then, that as long as some species are expanding their territories, the fragile competitive balances between other species will be temporary and local, partial exterminations frequent. So the uniform perpetuation of the present system of nature as currently observed includes "a principle of endless variation," a source, that is, of unending exchanges in the species that are present at any and every place on the globe.[25]

Now, if we go on to include in our predictions as to the future, and so also our retrodictions as to the past, our knowledge of geological changes now in progress, we can see that the present system must entail not merely local losses of ground and partial losses of numbers, but also total losses, extinctions of species. For geological changes affect the habitations and so disrupt the stations of species in two ways. They facilitate invasions by some species and they force retreats on others. Extinctions come when the invasions succeed and the retreats do not. Sometimes the invasions, retreats and so the extinctions can result simply from changes in land and sea themselves. New islands arising between old ones and a continent may allow mainland species to invade where they had not reached before. At other times the climate changes resulting from changes in land and sea will contribute. Such climate changes may be local or global. As the prevailing wind works on its sand the Sahara desert creeps eastward; many species adapted to moist, lowland stations will fail to move into suitable areas to the east, for their stations there will be occupied already by other species. Again, if a central part of the desert itself should be subjected to "the upheaving power of earthquakes" over an "immense series of ages" with repeated "volcanic eruptions" such as, in 1755, raised a mountain 1700 feet on the Mexican plateau, then eventually there would come effects in the resulting chain of snow-covered and lake-dotted mountains fatal to many species in the surrounding desert plains. As earthquakes and volcanic eruptions continued, "there would be floods, caused by the bursting of temporary lakes and by the melting of snows by lava." Moreover, these "inundations would deposit alluvial matter far and wide over the original sand at all levels," as the igneous upheavals and aqueous erosion combined to shape and reshape the country. Eventually, the Sahara, fertilized and irrigated by streams, and covered by jungles, would have lost its present species and be fit to receive others entirely different "in their forms, habits, and organization."[26] Existing causes may, indeed, effect even greater changes. For any elevations, subsidences, erosions, and depositions that bring over eons an overall movement of land away from the poles, and thereby cause a rise in the mean temperature of the whole earth, must also cause temperate species to move toward the poles, and arctic and antarctic species to face partial and total extinction; while the reverse tendency will see tropical species extinguished and replaced by temperate species in equatorial regions.[27]

Even the most widespread faunal and floral losses can be explained, therefore, by extrapolating from what has been observed in historic times. When, Lyell holds, we add together "what we have said of the habitations and stations of organic beings in general" and the effects of the "igneous and aqueous causes now in action," we see clearly that "species cannot be immortal, but must perish one after the other like the individuals which compose them."[28]

Such, then, is the account of extinction that Darwin was considering in February 1835 as he sailed up the west coast of South America and pondered his east coast finds once more. His decisive note is on three quarto sheets that in pen, ink, texture, and watermark (Collins 1828) are uniform with those of the *Diary of Geological Observations.* They have been cut with scissors and put, presumably by Darwin himself, with later notes for his book on South American geology. But this is no reason, nor is there any other, to doubt that the entries on them, including the footnotes, and also the reflections on an essay by Buckland that follow immediately, were indeed made, as the initial heading indicates, in February 1835, as part of that diary.[29]

We will do best as it turns out to have the full text quoted here, with Darwin's own parentheses as he has written them himself:

Feb 1835:
The position of the bones of Mastodon (?) at Port St. Julian is of interest, in as much as being subsequent to the re-modelling into step of what at first must especially appear the grand (so called) diluvial covering of Patagonia. —It is almost certain that the animal existed subsequently to the shells, which now are found on the coast. I say certain because the 250 and 350 ft plains must have been elevated into dry lands when these bones were covered up & on both these plains abundant shells are found. We hence are limited in any conjectures respecting any great [these last two words inserted later?] change of climate to account for its former subsistence & its present extirpation. In regard to the destruction of the (a) former large quadrupeds the supposition of a diluvial debacle seems [underlining made later] beautifully adapted to its explanation; in this case, however, if we limit ourselves to one such destructive flood, it will be better to retain it for the original spreading out of the Porphyry pebbles from the Andes. I have reason to suppose that this was anterior not only to this group of bones but also to the immense numbers buried in the Pampas. The arguments are drawn from the connection of the Tosca rock & matrix of the gravel at R. Negro; also the height of the plains.

The note (a) on the verso side reads:

It may well be expected if the remains [this last word inserted] large quadrupeds found in so many places in the Americas were overwhelmed by Debacles, with their bones immense quantities of trees (& stones if such existed) would be found. This is not the case in S America as far as I have seen.

The text of the next recto side (p. 2) proceeds:

With respect then to the death of species of Terrestrial Mammalia in the S part of S. America I am strongly inclined to reject the action of any sudden debacle. —Indeed the very numbers of the remains render it to me more probable that they (a) are owing to a succession of deaths, after the ordinary course of nature—as Mr. Lyell supposes species may perish as well as individuals; to the arguments he adduces I hope the Cavia of B. Blanca will be one more [this last word inserted] small instance of at least a relation of certain genera with certain districts of the earth. This correlation [this last word inserted] to my mind renders the gradual birth and death of species more probable. The large quadruped of St. Julian lived probably when the land was from 100 to 200 ft. lower than at present. If my belief [is granted] (it will hardly be granted by those who have not seen the plains of Patagonia) that the elevations of the whole coast have everywhere acted with remarkable equality of force; this same lowering would submerge the Pampas where many bones are found & afford a weak presumption that the animals to which these remains belong lived at same time. This reasoning will also include [p.

3 recto] the bones found by Humboldt in plains of Orinoco.—(V my index) Is the animal of St. Julian a Mastodon? Even if a change of climate could be granted it is scarcely possible to believe the plains of gravel ever could have supported a much more luxuriant vegetation. We must suppose like the Camel of Eastern climes or the Guanaco which now lives here, that it was fitted for a stunted vegetation.—Yet their whole genus is furnished with a short neck.—If a rapid river ever flowed into this bay the Carcase might have travelled from the slope of the Andes.—The latitude 49°-50° is certainly high. (corresponding to North of France & to Germany in Northern Hemisphere) for a large quadruped of this [roughly the bottom quarter of the page is here cut off].

The note (a) on the verso side of the previous page (p. 2) reads:

(a) The following [this last word inserted] analogy I am aware is a false one; but when I consider the enormous extension of life of an individual plant seen in grafting of an Apple tree & that all these thousand trees are subject to the duration of life which one bud contained. I cannot see much difficulty in believing a similar duration might be propagated with true generation—if the existence [these two words inserted to replace "gradual death"] of species is allowed, each according to its kind, we must suppose deaths to follow at [last word replaces "one after"] different epochs, & then successive births must repeople the globe or the number of its inhabitants has [last word inserted] varied exceedingly at different periods.—A supposition [this last word replaces "fact"] in contradiction to the fitness which the Author of Nature has now established.[30]

When we compare it with parallel texts in the *Journal of Researches* and the *Geological Observations,* the argument of Darwin's first paragraph is easy enough to discern as it moves from the position of the Port St. Julian bones to the difficulties for either a diluvial or a climatic explanation for extinction in all these mammalian cases.[31] The bones, putatively of a mastodon, Darwin found embedded in a mass of reddish loam soil filling a gully once worn through part of a cliff on the south side of the harbor at Port St. Julian. On the plain at the top of the cliff he had found, as nowhere else he had seen, a thin covering of this loam soil over the uppermost gravel. So the animal must have lived after this plain, subsequently elevated to its present level, first emerged from the sea. The animal must then have lived after the higher plains, now 250 and 350 feet above the sea, were elevated even earlier from the ocean bed to form dry land. So it must have lived on coastal plains and become extinct after the mollusk species now living in the ocean first originated. For both those more elevated plains have fossils of these very same mollusk species scattered on their surface. We cannot therefore suppose that it could flourish there, then, and not now, because the climate was previously very different, for had the ocean changed its temperature other mollusk species would have invaded the region. Nor can we suppose that a change in climate brought on its extinction, for the small elevationary steps would entail no such change.

A diluvial debacle could have extinguished this mastodon and those other quadruped species without a climate change; but, Darwin reasoned, the only flood independently evidenced by the rocks in Patagonia and in the Pampas area to the north is needed to explain the spreading of the porphyry pebbles in the gravel beds, and so it probably came before the mammals were living in the region; for the River Negro valley sections, and the sameness of level in the Tosca rock beds of the Patagonian and Pampas plains, show that these pebbles were probably deposited prior to depositions of Tosca rock that took place before there was formed the land the mammals occupied. Against a flood, too, as a cause of extinction is the lack of trees and stones buried with the large quadruped bones, at least in South America.

This conclusion, against both diluvial extinctions and extinctions from changes of climate, Darwin defends further in his "Reflections on Dr. Buckland's beautiful paper on the fossil remains of large Quadrupeds in N.W. Coast of America." By referring the formation of the coastal mountains of that region, also, to many small elevations, and the various lowland formations to ocean or estuary rather than diluvial depositions, Darwin directly counters Buckland's own conclusion that flooding and cooling of large regions must have been responsible for the North American remains just as they were, in Buckland's judgment, for the more famous Siberian mammoth remains.[32]

It is thus fairly clear how Darwin has reasoned, from the position of the Port St. Julian and other bones, to his conclusions about the extinctions of these species. But his subsequent reasoning is much harder to establish with precision and confidence, his reasoning, that is, *from* the great numbers of the remains and *from* the animals often belonging to genera whose extant representatives are today confined to that region, *to* confirmation of Lyell's gradual, successive exchange of species. Most likely, his thought is that the Port St. Julian find has shown that these animals, like those now found as fossils in the Pampas to the north, lived on these plains when the land was 100 or 200 feet lower than now, and that their carcasses were carried by small or at least slow rivers, which often deposited them close to the coast at that time. The great numbers of the remains would thus confirm this view, presumably, because if the carcasses or skeletons were brought from a long way off they would be more likely to be deposited less densely and over a wider area of ocean bed. So there is additional support for the conclusion that little change in land, sea, climate, or consequently, vegetation has occurred since these animals flourished here. Moreover, the continued representation up to today of certain genera such as that of the cavies (harelike animals) peculiar to this region, again suggests that little change in conditions has intervened. Hence these extinctions seem to have occurred not during a special, unusually active period of land, sea, and climate change unlike the present, but on the contrary while everything else was going on very much as it is, observably, or at least presumably, now: a conclusion suggesting in turn that these extinctions of earlier species and compensating origins of similar, extant ones were, as Lyell would insist, events belonging to the present system of nature.

Now although we cannot be sure that this was exactly Darwin's reasoning in this passage, his footnote to it makes very plain where this Lyellian line of thought had taken him. He has found no changes of conditions that could plausibly cause these successive extinctions, not even such minute changes as Lyell himself would require to bring local disturbances in the finely balanced economy of nature. So, Darwin has now to rethink the whole problem of extinction; for his Lyellian explanation of the region's history has made even a Lyellian explanation of its quadruped inhabitants' extinctions implausible.

## 5. The Propagation of Trees and Zoophytes and the Common Laws of Associated Life, Death, and Extinction

The brief note to his second page introduces an analogy that—despite any initial, momentary qualm he may or may not have sincerely felt about its being a false one—was to dominate all his thinking about extinction for the next three years, and so influence, indirectly but none the less fundamentally, all his thinking about life and its laws for the rest of his days. Its introduction marks a truly consequential moment in Darwin's own life. We need, then, to be clear as to how the analogy works, and how it was worked into a theory of species change.

Consider, first, the grafting process Darwin cites. From parallel texts, especially in Lyell's *Principles* and in Darwin's own *Journal of Researches,* that we will examine in due course, we can see that Darwin is thinking of the technique whereby shoots (buds in his word) are cut from a prized apple tree and grafted onto other seedling trees, which have had their stems cut back in readiness to receive the grafted shoots. After these shoots have grown, more shoots can be taken from them and grafted onto stems on still other trees. In this way, a hundred or indeed a thousand graft trees can be had from the single bud or shoot that the prized tree originally was before it began its own growth as a seedling. So, the commonly reported finding decisive for Darwin's analogy is that this process cannot be repeated indefinitely because after a certain period of time all the material, the hundreds of trees tracing to that first tree and, so, to the single shoot or bud that it once was, simply die all of a sudden. It is as though that bud and all its successors formed but one extended tree which dies, in due course, of old age. What the grafting achieves, then, is a great extension, but a strictly limited extension in years, of the life of the original, individual tree. And what Darwin's analogy conjectures is that, as in grafting, so in "true generation," that is sexual reproduction, there might be passed on a similar limitation, of the extended life given to the first members of the species in the production of the subsequent generations that are produced sexually.

Now, this conjecture would entail a successive exchange of old species for new if supplemented by two additional hypotheses or suppositions. First, Darwin supposed that "the existence of species," that is the total, limited, time in years that any species lasts, depends on its distinctive constitution and specific characters and so varies from one species to another. The extinctions of the various species extant at any time will, therefore, not come all at once but at different epochs. But has the globe, then, steadily lost species, and has it had, consequently, very different numbers of species at different times? Or should we suppose that origins, births of new species are constantly compensating for these successive losses? At the present time we observe that God, working through nature, ensures a constant conservation of the planet's overall fitness for life; so we have no ground for supposing that its fitness for life, or therefore its fullness with life, its total number of species, has declined from ages past.

This teleological and theological supposition obviously takes its precedent and its legitimacy for Darwin from the *Principles.*[33] But, in fact, this is no less true of every other word and phrase in the tree analogy note. Take the interpretation of grafting itself. Lyell had discussed whether the production of domestic varieties of animals and plants, by farmers and gardeners, supported Lamarck's view that wild species are indefinitely mutable in changing conditions. He had argued that men can sometimes conserve a novel variation in a plant by means never used by nature in the wild; but that these means of vegetative propagation only perpetuate the new variety for a limited time, because the extended life of the individual tree achieved by such means is itself strictly limited. "The propagation of a plant by buds or grafts, and by cuttings," is not employed by nature, he wrote. And "this multiplication," like that "produced by roots and layers" which is found naturally, seems only to "operate as an extension of the life of an individual, and not as a reproduction of the species such as happens by seed." Citing botanical authorities, Lyell insisted that all plants increased by layering or grafting keep "precisely the peculiar qualities of the individual to which they owe their origin," and that "like an individual they have only a determinate existence," in other words a predetermined span of life in years, which is "in some cases longer, and in other shorter." Horticulturalists admit, therefore, that no such garden varieties are "strictly permanent," for they "wear out after a time," so that sexual generation by seed has to be resorted to eventually; but then they find that there is such a strong "tendency in the seedlings to revert to the original type"—so many of these sexual seedlings, that is, have the old and not the new cherished characters—that often they cannot "recover the desired variety."[34]

Moreover, Lyell had discussed respectfully, before of course finally rejecting it, the conjecture that the extinction of species could be explained as due to an analogous predetermined wearing out of a constitutionally limited life. Immediately after first raising questions about the duration and successive extinction of species, and immediately before introducing his own arguments for the sufficiency of what is known of the agencies affecting species stations to provide answers to them, Lyell had discussed approvingly how the Italian geologist Brocchi had sought to explain the extinction of many Mediterranean mollusk species. Rather than "seeking a solution to this problem, like some other geologists of his time, in a violent and general catastrophe," he had tried, Lyell said, to "imagine some regular and constant law by which species might be made to disappear from the earth gradually and in succession."[35]

According to Brocchi, Lyell said, the death of a species might depend, like that of individuals, "on certain peculiarities of constitution conferred on them at their birth"; and as the longevity of the one depends on the weakening of its vital force, the duration of the other might be determined by a power of proliferation that eventually weakens, so that there comes a time when the new embryos cannot grow and mature, but die and so bring the whole species to its death. Now, says Lyell, one could well agree with Brocchi, "as to the gradual extinction of the species one after another by the operation of regular and constant causes, without admitting an inherent principle of deterioration in their physiological attributes." Brocchi must have known of extrinsic causes, such as climate changes and increases in human and other populations, which "might gradually lead to the extirpation of a particular species" although its fecundity stayed as high as ever. So, if there are such "known causes," present in "the vicissitudes of the animate and inanimate world," which might be capable of producing "the decline and extirpation of species," then Brocchi should have investigated all their possible effects "before he speculated on any cause of so purely hypothetical a kind as the diminution of the prolific virtue."[36]

This methodological censure is unequivocal, but it provides for a reversal of its argument if and when appropriate factual findings were to be made. And Lyell's discussion ends by indicating what findings would make this weakening of vital powers no longer a purely hypothetical but an acceptably evidenced hypothetical tendency.

> If it could have been shown that some wild plant had insensibly dwindled away and died out, as sometimes happens to cultivated varieties propagated by cuttings, even though climate, soil, and every other circumstance should continue identically the same—if any animal had perished while the physical condition of

the earth, the number and force of its foes, with every other extrinsic cause, remained unaltered, then might we have some ground for suspecting that the infirmities of age creep on as naturally on species as upon individuals.[37]

But, insists Lyell, one must respect the principle that causes already known to be adequate are always preferable in hypotheses, so "in the absence of such observations," he will now see whether in the changes wrought in stations there is "a cause fully adequate to explain the phenomena" of extinction.[38]

We can see, then, that what Darwin did—on pondering the extinctions of his large quadrupeds, and on deciding that these events could not be explained by Lyell's own ecological and climatic theory—was simply to go back to the start of Lyell's development of that theory and adopt, as fitting the anomalous facts much better, the only theory that Lyell had judged potentially consistent with the proper principles of hypothesizing in geology.

So, Darwin's initial adoption of his Brocchian hypothesis is fairly fully explained. But, in order to understand his subsequent elaboration of it, we have to appreciate that he was, in 1835, not only a student of Lyell but a former student of Robert Grant, his principal teacher in zoology at Edinburgh. For it was under Grant's formal and informal instruction that Darwin had first become fascinated by the various modes of generation found in "zoophytes": plantlike animals such as the colonial polyps of the genus *Flustra,* a genus usually included today in the phylum *Bryozoa,* whose classification was subject to prolonged discussion and disagreement in Grant's time.[39]

A full account of Darwin's development as an invertebrate zoologist would have to follow him from his first apprenticeship to Grant, through the *Beagle* years, when records and speculations concerning colonial zoophytes often dominated, especially early on, hundreds of pages in his *Diary of Zoological Observations.* It would then have to follow him on to the barnacle studies where, once again, difficulties about modes and means of generation raised many of the principal problems. But for our purposes here we need concentrate only on the two main issues Darwin himself took up at Edinburgh, in a notebook that still survives, and made central to the paper he helped Grant prepare on the *Flustrae.* First there were questions about the extent of the independence of the more or less distinct and separate individual organisms making up the colonies arising by successive buddings from founding individuals. Second, there were questions as to whether the small independently motile polyps produced by some species were involved in sexual or in asexual generation. The two issues were thus related, for greater mutual dependence in the individual buds suggested that the whole colony be reckoned a single organism analogous to a tree, while clear signs of independence and motility in asexual parts or stages weighed against the affinity with plants. Darwin's own observations at Edinburgh, as he and Grant agreed, seemed to settle these questions in favor of the *Flustrae* and similar genera being animals.[40]

These themes are indispensable for our understanding of Darwin in the years from Edinburgh to the Andes, and so, too, in the years from the Andes to Down House. In the publications relating to the voyage, Darwin may well seem, as a zoologist, most at home with birds and mammals. But that was not always so. Long before he had much confidence in himself as a student of higher animals, his main expertise as a zoological observer and theorist was with these zoophytes, or Polyzoa as they were also called, and with the problems posed by their propagation. In the last eighteen months of the voyage, as in the years before, his zoological diary returns again and again to these colonial or "compound" animals, as he calls them, and their vegetative asexual generation. His interest in trees and their propagation by cutting is also shown in notes on observations made in Chiloe, where thick branches of orchard trees rather than small stems were used as cuttings to be planted, ungrafted, into the ground.[41]

His geological diary does not take up explicitly his new, Brocchian explanation for extinction. It does, however, abound with evidence that he did not give up the views on South American geology that it presupposed. These views were essentially unchanged when he presented them to the Geological Society of London in the Spring of 1837.[42] The next explicit elaboration of that extinction theory in any documents extant today seems in fact to have been made earlier that year. For we find it in what Darwin dubbed his "Red Notebook," in entries made most probably in February or March of 1837, and so just two years after his first formulation of the theory.[43]

Darwin's analogy in his February, 1835 note depended on arguing for a common feature—the transmission of a limit on the duration of life—in two forms of generation (the sexual reproduction of quadruped mammals and the vegetative propagation by cutting of a fruit tree) that would normally be considered as very much contrasting. Now, this analogical argument could be strengthened if Darwin could show that all forms of generation shared some one fundamental feature and that, as variations on this fundamental theme, they differed only in degree not kind. This proposal in turn could be supported by finding forms of generation that were, with respect to this fundamental feature, intermediate between the sexual reproduction of higher animals and the vegetative propagation of trees by cuttings.

In a note in the Red Notebook, Darwin adopts this line of thought. "Propagation," he reflects, "whether ordinary hermaphrodite or by cutting an animal in two (gemmiparous by nature or by accident)" always shows us "an individual divided either at one moment or through a lapse of ages.'' In a word, propagation or generation consists in essence of a single original individual being divided either over time or all at once. So, we should not be surprised to find cases where the division does not go to completion and where the products of the division stay partly united in a colony, nor to find this incomplete division in animals which also show other kinds of generation, other divisions in which, as in ordinary sexual generation, the products become quite separate: ''Therefore we are not so much surprised at seeing Zoophite producing distinct animals, still partly united & egg which becomes quite separate.'' In conclusion then: ''Considering all individuals of all species [Darwin means of any single species] as each one individual divided by different methods, associated life [i.e., any colonial life such as we see in polyps] only adds one other method where the division is not perfect.'' So, conversely, as Darwin is ultimately concerned to argue, we can point to the many cases of associated life among plants and animals as evidence that this understanding of generation as division, whether successive or simultaneous, whether total or partial, is a correct one.[44]

Two passages in the *Journal of Researches* that, with the possible exception of a footnote, we have no reason to doubt were written between February and June 1837, show us, when taken together, that in the spring months of 1837 Darwin was very much concerned to develop his case for this view of generation and for its application to extinction. In the first, after reviewing his observations on *Flustra* and similar genera, he argues for comparing the plantlike growth of a polyp colony with the ''known organization of a tree,'' with, that is, the accepted view of a tree as being a colony of individual buds produced by successive divisions starting from a single bud. In the other passage, which actually comes earlier in the book, he had shown how difficult it would be to reconcile many extinct quadruped finds, not only in South America but all over the world, with the Lyellian view that ''such simple relations, as variation of climate and food, or introduction of enemies, or the increased numbers of other species,'' are the ''cause of the succession of races.'' He stresses now that there are several cases of a single species, like the mammoth, becoming extinct apparently very quickly over a very wide range including more than one continent. If any ecological changes were responsible, therefore, ''they must have been changes common to the whole world; such as gradual refrigeration, whether from modifications of physical geography, or from central cooling.'' But that assumption runs straight into the difficulty that ''these supposed changes, although scarcely sufficient to affect molluscous animals either in Europe or South America, yet destroyed many quadrupeds in regions now characterized by *frigid, temperate* and *warm* climates!''[45]

Now, these cases of extinction, says Darwin, recall forcibly ''the idea (I do not want to draw any close analogy) of certain fruit-trees'' that, although "grafted on young stems, planted in various situations, and fertilized by the richest manures, yet at one period, have all withered away and perished." In such a tree propagation, he proposes, we see that a "fixed and determinate length of life has been given to thousands and thousands of buds," even though they have been produced not all at once but "in long successions." Likewise, then, in a species of animal we may have a long succession of "individual germs," equivalent to successive individual buds in plants, with a no less determinate total life span. In most animals, to be sure, "each individual appears nearly independent of its kind," capable, that is, of living without other members of its species; but all the members of a species may be, nevertheless, "bound together by common laws" as are "a certain number of individual buds in the tree, or polypi in the Zoophyte."[46]

Darwin's preoccupation with extinction, with finding generational rather than ecological causes for it, lasted, then, well into 1837; and so well into the time when he was actively favoring the view that species' origins are brought about by the transmutations of earlier species. To appreciate the significance of this overlap, however, we need to prepare the ground, by looking first at Lyell on species' origins and then at Darwin's break with Lyell on that issue.

## 6. Lyell and the Laws Governing the Animate Part of the Terrestrial System

We need to drop back next to consider how Lyell's explanatory program, integrating Huttonian geology (as revised by him) and Linnaean ecology (as refined by De Candolle), was elaborated in answering the two earlier questions—on the duration of species and on the singularity or otherwise of their origins.

To establish that the hypothesis of single original stocks, and so of singular geographical origins, is consistent with what we observe today, Lyell asks the reader to consider a vast imaginary experiment.

Suppose all life is removed from the Western Hemisphere; and suppose naturalists knowledgeable of their stations were to repopulate it with Eastern species, one pair of each species only to be introduced. Three consequences follow without our considering geological changes at all. First, physical differences in stations, differences of soil and climate especially, would require species introductions to be made at different places; while further, organic, differences in stations, differences in food, hosts, and so on, would require the species to be introduced successively over a long span of years, because many species would depend for their livelihood, directly or indirectly, on a prior growth of population in others. Second, thanks to physical and, later on, organic barriers to migration and dispersion, there would result "distinct botanical and zoological provinces," areas containing many species found nowhere else; for many of the species introduced in the temperate southern part of South America, for example, could not spread east, west, or south because of ocean; and blocked to the immediate north by tropics, they would be missing from temperate areas in the far north that could well suit them if only they could get there. Third, there would be some small areas, surrounded by barriers, that would appear to have had more than their share of species introductions. But, like those "*centres* or *foci* of creation" in the world we observe, which seem to have seen unusually many origins of species, in fact they need not have done. For, if any confining area, like an island, has lasted long enough, it will have had originate in it many species that have remained there, while the equally many species originating in equal areas beyond the barriers will, as in the surrounding ocean, have been dispersing themselves far and wide.[47]

If geological changes are now added to this single stock hypothesis, it can, Lyell argues, yield a fourth consequence: the existence of real, not merely apparent centers of dispersion, areas from which many species have dispersed into surrounding regions. For, consider a confining area, like an old, solitary ocean island, that has had several species originate on it; such an area may eventually supply other areas, by migration, with various selections from among its native species when geological changes later make those areas suitable and accessible, as would happen, for example, were new volcanic islands to arise in the ocean near the old one.[48]

Now, the addition of geological changes can help Lyell's single stock hypothesis resolve many explanatory challenges; but in doing so it raises a further, general problem. For, if geological changes facilitating migrations have gone on throughout the long period since the oldest extant species first originated, would we still expect to find any distinct botanical and zoölogical provinces? Or, rather, conversely, since we know that important geological changes have occurred in this long period, how can we reconcile that fact with the existence, today, of these distinct provinces, these areas with many species found nowhere else?[49]

Such apparent difficulties, in conjoining the single stock hypothesis and continuous geological change, can only be resolved, Lyell declares, when he has discussed not only the laws regulating the introduction of new species but also those which "may limit their *duration* on the earth."[50] Stations, habitations, and the continual changes consequent on constant igneous and aqueous action suffice here. For the adaptive fits of species to their stations being both close and complex, the stations themselves are so fragile that they can last only a limited time, since geological change must always be eliminating old ones and creating new ones. This elimination of stations and so species is gradual in both senses; it is gradual, collectively speaking, for the species extant at any time become extinct successively. And it is gradual for each species, because before finally becoming extinct, its range and its numbers will be reduced and confined. So, some species are rare and local because recently originated, others because the causes that will eventually extinguish them altogether have already done so partially. That many individual species are local is thus no longer anomalous.[51]

Moreover, if we consider the effects of continuous geological change—throughout the vast period in which the species now extant have been originating and replacing others as they became extinct—then we can also resolve those difficulties zoologists and botanists have had in relating distinct faunal and floral provinces to features in the earth's physical geography. For some people, Lyell says, have supposed mountains to be "centres of creation" from which have radiated many species now spread over whole continents, while others have objected that mountains are often, rather, barriers between two distinct provinces. The Lyellian geologist can "reconcile both these theories," however; for any chain of mountains, which has arisen within the period when extant species began to originate, could well have been a center from which species have since dispersed into more recently elevated lowland plains; while any mountains older than the oldest extant species will now divide one province from another.[52] Lyell's gradual species extinctions and origins and gradual geological changes can, then, he argued, provide a sound foundation for integrating geographical knowledge, about the present distribution of species, with generalizations about the locations, the distributions, of species at their origins and about their dispersions and migrations from those origins.

Such an integration of geology and geography may seem so obvious, and anyway so familiar, from Darwin's and Wallace's writings two decades later, that it will be hard to think of it as a revolutionary proposal. But, as a self-conscious innovation in the science of animal and plant geography, it was indeed new and radically so. In fact, unlike Lyell's proposals for reforming geology itself, it was very successful, too successful for his own good, or at least the good of his name; for the simple reason that it was so thoroughly adopted by younger men like Forbes, Hooker, Gray, and Wallace that—as Lyell, ever jealous of his priorities, was keenly aware—his name was not always explicitly attached to it. Thus Forbes in 1846 was already taking it that everyone knew that it traced to Lyell. And Wallace's 1855 paper "On the Law Which Has Regulated the Introduction of New Species" not only took its title from Lyell quite silently, it began defining its very objectives in three opening paragraphs that insisted on the "light thrown" on geographical distribution by "geological investigations"; the three paragraphs being nothing more nor less than an encapsulation of Lyell's teaching, but with no reference to the *Principles,* which, Wallace clearly assumed, would be instantly recognized as this source.[53]

So Lyell's biogeographical revolution is a little one only because (for those who subscribed to it) it transformed a rather minor if not marginal science—animal and plant geography, or zoological and botanical geography (biogeography as we shall allow ourselves to say hereafter). It was a major revolution in a minor science.[54]

The revolution itself is, nonetheless, of prime significance for Darwin's science, and for one reason, especially, that will be worth mentioning here, even though it must be dwelled on again later. A question is sometimes raised as to how Darwin's thinking, so "historical" as it is, could be indebted to Lyell's, so "ahistorical" as that has seemed to some readers to be.[55] The paradox is a suspect one, trading as it does on a vague and impressionistic contrast drawn with troublesome words. But even on its own questionable terms, it is easily resolvable once we concentrate our minds on what Lyell did to biogeography.

Lyell took over from older authors, especially Linnaeus as refined by Prichard and De Candolle, a concept of individual species that was in obvious respects historical and geographical: each individual species has had a spatially and temporally unique origin and a subsequent career that has included multiplication in numbers, expansion in range and limited diversification in character. It was, thus, in being historical and geographical also genealogical. What Lyell did was to make the understanding of species origins historical and geographical not merely when they were considered individually but also, as in explaining the spatial and temporal representation of supraspecific groups, when considered collectively; while, of course, explicitly rejecting the proposal (which he read, quite mistakenly, into Lamarck) that the understanding of species origins should also be genealogical when taken collectively.

It is then ironic, not to say paradoxical, that people have sensed any paradox in the debt of that "historical" thinker, Darwin, to the "ahistorical" geologist, Lyell. For in biogeography Lyell was seen in his own generation as the great historicalizer, if such a word be allowable. Of course, there is more, far more, to Darwin's thinking than the mere addition of genealogy to the Lyellian history and geography for supraspecific groups. But we will not understand what more there is unless we first understand how that addition was initially made and seen by Darwin as a completion of the historicalization of biogeography started by Lyell.

Let us next, then, make quite sure that we have a firm grasp of what it was that historicalization included and how it was related to the wider aims of the *Principles.*

Counting himself fortunate to have met De Candolle in person, Lyell greatly admired the Genevan's essay of 1820 on botanical geography. He had worked out his own biogeographical theorizing very much as a response to the state of the science as expounded in that essay. In 1829, after a detour to Geneva to "get some ideas from De Candolle for my new geologico-botanical theory," he told his father that geology is "destined" to throw "much light" on botanical geography and receive much from it "in return." By the "mutual aid" of the two sciences "we shall very soon resolve," he declared, "the grand problem, whether the various living organic species came into being gradually and singly in insulated spots, or centres of creation, or in various places at once, all at the same time." He was already convinced that the latter was false and proceeded to explain how the first hypothesis, if combined with his views of recent changes in the organic world, can account for facts otherwise mysterious. Sicily, he said, naming what would become his favorite instance of his new theorizing, has puzzled people by seeming to have "much less than its share of peculiar indigenous species"; but "this should be"; for "I can show that three-fourths of this isle were covered by the sea down to the period when nine-tenths of the present species of shells and corals (and by inference plants) were already in existence."[56]

Years later he would write on his first edition that it included "the first attempt" as far as he knew, to show "how former changes in the geography and local climate of many parts of the globe must be taken into account" when one tries to explain the present "provinces of plants"; the "changes alluded to having been proved by geological evidence to be subsequent to the creation of a great proportion of the species now living, and these having been . . . introduced in succession and not all at one epoch." And he cited his elucidation of the Sicilian fauna and flora as a telling case in point.[57]

De Candolle had proposed, as his leading hypothesis, that the stations of plants are "uniquely determined by physical causes acting presently [*actuellement*]" while the "habitations may well have been in part determined by some geological causes which no longer exist today." And this hypothesis, he said, easily explains "why certain plants are never found wild in some places where they flourish as soon as one transports them there." But, he admitted, the hypothesis shares "in the uncertainty of all ideas concerning the former state of our globe and the first origin of organised beings." To illustrate this uncertainty introduced by all such ideas, he cited the difficulties surrounding the conjecture of the eminent Willdenow and others: that mountains, being the lands "first uncovered by the waters," must have been the primitive centers from which many plant species originally dispersed. That notion, says De Candolle, fits well enough the mere fact of floral provinces; but it is hard to square both with the temperature differences between plains and mountains and with those cases where a mountain chain serves as a limit rather than a center of a distinctive vegetation.[58]

When De Candolle came to conclude his whole discussion, then, he emphasized how actualistic were the laws of physiology and physics, the laws that allow the science of geography to be distinguished from the geologist's unactualistic quest for other laws relating to the unknown origins of things. Reviewing the evidence for new species arising either by the transmutation or by the hybridization of older ones, he decided that it is inadequate; and that "there exist in organised beings permanent differences which cannot be attributed to any present causes of variations." These differences, he says, constitute species. And these species are distributed over the globe partly "according to laws" which one can infer directly "from the combination of known laws of physiology and of physics," and partly "according to laws which appear to relate to the origin of things and which are unknown to us." And that, he declares, "is the point where botanical geography is obliged to stop."[59]

We have already seen how Lyell's actualistic and uniformitarian understanding of geology led him to insist that this point is precisely the one where his own science joins forces with biogeography. But, before analysing further Lyell's integration of geography and geology, we may note in passing that the position of the other doyen of botanical geography, Humboldt, was in one decisive respect the same as De Candolle's; for he too wanted biogeography kept separate from geology, and so free from the uncertainties that any historical inquiry introduces into any geographical science. This is not the place to assess the many ways Humboldt may have influenced Darwin. But it may be remarked that one suggestion of Cannon's is plainly incredible: namely, the proposal that, insofar as Darwin was peculiarly preoccupied with the constantly changing biotopography (Cannon's word) of the earth's surface, he was indebted to Humboldt rather than Lyell.[60] The suggestion is doubly implausible: first, because Lyell is always identified by Darwin himself as the source for his interest in bringing geology to bear on biogeography; and second, because Humboldt (even as late as the 1840s when he was writing *Kosmos*) always favored a contrast—which he knew echoed a famous passage in Plato's *Timaeus*—between geography, as certain knowledge of the ways things are, and geology as uncertain conjecture about how they could have come to be that way.[61] Nor should Humboldt's favoring of such a contrast be surprising. Following the tradition founded by Werner, he distinguished, in geology itself, between "geognostical" certainties about the present superpositional relations among strata, as observed today, and "geological" speculations as to how those relations may have arisen in the original deposition of those strata. Nor should Darwin's failure to follow Humboldt—in segregating geology from geography and geognosy from geology—be surprising. Lyell had explicitly rejected both segregations and made it plain that no one accepting them could count himself committed to the reinterpretation, offered in the *Principles,* of what geological science was all about.[62]

On historical biogeography itself, we have already met with Lyell's decisive innovation: the successive, one-by-one introduction of the species extant, today, over a vast period that has seen major changes, gradually wrought, in land, sea, and climate all over the globe. It was this innovation that entailed directly, as Lyell emphasized, that the number of species common to both sides of a barrier, such as a mountain range or stretch of sea, will have been determined by the age of the barrier. Consider a young barrier that has therefore arisen recently: few of the species formerly living throughout the region that it now divides will have become extinct. Few then will have been replaced on either side of the barrier by new species. There will then be few species peculiar to each side. Consider by contrast an old barrier; there will have been many extinctions among the species formerly on both sides of it, and so many new species will have originated since, each confined to one side only.

The reasoning is such an obvious corollary of Lyell's general thesis—of gradual changes in physical geography and successive exchanges of species—that it is easy to forget that no one could accept it who had not first embraced the highly controversial general thesis that it presupposes.

It is easy, too, to overlook a further corollary of this reasoning itself. While the number of indigenous species peculiar to each side of some barrier can be referred to the age of the barrier, the representation there of genera, families, and orders cannot be. For that supraspecific group representation is of course determined by the characters of the new species originating there.

We have already seen what Lyell's answer to this further question is, however: adaptation. He took it as obvious that if the conditions on either side of the barrier are similar, then they will require similar (congeneric or cofamilial) species; if different, then species of distinctive genera or even families will be called for. In 1853 he summarized his first edition position crisply and carefully:

> From the earliest period at which plants and animals can be proved to have existed, there has been a continual change going on in the position of land and sea, accompanied by great fluctuations of climate. To these ever-varying geographical and climatal conditions the state of the animate world has been unceasingly adapted. No satisfactory proof has yet been discovered of the gradual passage of the earth from a chaotic to a more habitable state, nor of any law of progressive development governing the extinction and renovation of species, and causing the fauna and flora to pass from an embryonic to a more perfect condition, from a simple to a more complex organization.
>
> The principle of adaptation to which I have alluded, appears to have been analogous to that which now peoples the arctic, temperate, and tropical regions contemporaneously with distinct assemblages of species and genera, or which, independently of mere temperature, gives rise to a predominance of the marsupial or didelphous tribe of quadrupeds in Australia, of the placental or monodelphous tribe in Asia and Europe, or which causes a profusion of reptiles without mammalia in the Galapagos Archipelago, and of mammalia without reptiles in Greenland.[63]

If we ask then whether there was a theory of the origin of species in Lyell, in the sense of a theory as to what one would see if one were watching a species originating, the answer is that there is none. But in another sense there is a theory, indeed several theories, several explicit generalizations about the origins of species. Among them, it will be needless now to emphasize, was a theory as to what has determined the timing and placing of species origins; that is, a theory as to what determined that the marsupial species now extant in Australia originated when and where they did rather than somewhere else and earlier or later. And that theory was the principle of adaptation.

So, in considering the break that any follower of Lyell made with him over species origins, we have to consider which of the several generalizations about species origins the dissident protégé was rejecting and why; and which he was not and why.

Equally, in understanding the dissident protégé in relation to his mentor, we must distinguish individual theories from wider programmatic aims. Lyell brought a revolution to the science of biogeography; but his aims were never those of a biogeographer as such; he was not out to explain the geographical distribution of any particular fauna in order to have such an explanation for its own sake. His aims, again needless now to reiterate, were to use three evidential resources in establishing what he saw as a necessary prolegomenon to any attempts by geologists to reconstruct the history of any particular region, formation, or landscape. That is, he sought to find—in the changes falling within the human historical period, and in the geographical distribution of extant species, and in the stratigraphical distribution (in the Tertiary ages) of extant species and their extinct contemporaries—indications of the laws governing the animate part of the terrestrial system; laws, that is, about species, especially the timing, placing, and causing of their coming and going in the long run. That this was a new programmatic aim, and why, should now be plain; it presupposed his own new understanding of the problems posed by organic diversity.

## 7. Darwin, Adaptation, and the Characters of New Species

We are now in a position to appreciate what was involved in Darwin's first major break with Lyell over the origin of species; the one, of course, whereby he adopted the view that certain phenomena can be explained only by supposing new species to have arisen in the transmutation of earlier ones; with the corollary that species are not strictly limited in their mutability in changing conditions, but can sometimes become altered sufficiently for several distinct species to be descended from a single ancestral species.

Now, we do not have any documentation directly establishing either the date or the rationale for this development in Darwin's theorizing. However, it will be argued here that all the relevant circumstantial evidence leaves little doubt as to when and why it took place. For a case will be made for thinking that it came some time in 1836, most probably around the middle of that year and so in the closing months of the voyage, which ended in October. And a case will be made for its coming from Darwin's then deciding that he needed the transmutation of species as a solution to a particular problem in Lyellian historical biogeography: the problem of explaining why the new species originating on new land have often been very similar to other species already living on older nearby land, even when that land was quite dissimilar in climatic and other conditions. For this similarity of the new species to the old cannot be credited solely to adaptation. It should be ascribed to heredity as well; the characters of the new species being due both to their adaptation to the conditions that they are originating in and to their descent from earlier species in the same area.

In making the case for this interpretation of this break of Darwin's with Lyell, we do best to begin by looking at Darwin's views in the spring of the following year, 1837. For this turns out to be one of those occasions when the obligation to chronology is served most effectively by preparing the way for a chronological reconstruction (in the next section) with an analysis (in this section) that begins with the later phases of the narrative. Far from subverting our commitment to keep anachronism to a minimum, this preliminary analysis will thus eventually enable us to honor it all the better.

To propose that there is a pretty conclusive evidence, albeit circumstantial, for this date and rationale for Darwin's initial commitment to species origins by descent or transmutation is to depart somewhat from the consensus current among early Darwin specialists. But it is not to depart very far. For it seems quite widely agreed that Darwin was favoring transmutation very early in 1837, at the latest, and that some such problem in biogeography was then a prime consideration in his doing so.[64] So, what needs to be clarified here is the way the documentary evidence allows us to identify this particular problem as the initially decisive one, and to see why it was decisive several months before the Red Notebook entries (most probably in March, 1837), where it is very obviously prominent.

Before embarking on this clarification, however, it should be acknowledged that the reader may well wonder whether pursuing this matter can do anything for us, in analyzing any larger issues about the nature, sources, and development of Darwin's science, and whether the chronological and other details of this break of Darwin's with Lyell are worth pursuing further than specialist scholarship has pursued them already, which is after all pretty far. To this scepticism, one must reply that it is eventually worth it, because those details can bring with them sizable dividends for our understanding of how Darwin's theorizing proceeded over the next two years, in a series of shifting relations with Lyell's views; and because those dividends can in turn benefit our understanding of such topics as Darwin on adaptation and on progress, and Darwin in relation to Lamarck and to Erasmus Darwin.

It has long been customary to start discussions of the documentary evidence bearing on the present chronological questions with a famous entry for 1837 in Darwin's personal pocket journal or diary: "In July opened first note Book on 'Transmutation of Species.'—Had been greatly struck from about Month of previous March on character of S. American fossils—& species on Galapagos Archipelago. These facts origin (especially latter) of all my views." However, these words turn out to be very difficult to interpret with precision and confidence. Of course, it is plain enough that Darwin is here recalling March 1837 and the opening in July of that year of Notebook B (the alphabetical designations of the notebooks being his own). But, beyond that, which of course we knew already, so to speak, what is being recalled? In answering this, we must start by acknowledging that we do not know when Darwin wrote these words. The little diary or journal that they are in was not opened, apparently, until August 1838; but this entry seems not to have been made when others on that page were. It could, then, have been made more, much more, than a year after the events in question.[65]

Consider next the opening sentence. How is the phrase in quotation marks to be taken? Is it simply identifying the notebook as one concerned with the issue of fixity versus mutability of species? We cannot assume so, for two reaons. First, that notebook is in fact concerned, as Darwin surely still remembered, with a much wider range of topics than that particular issue. Second, as here written in quotation marks, it could well be referring not directly to that particular issue in itself, but to all those issues discussed under that heading in the opening two thirds of Lyell's opening chapter of his second volume; where Lyell discusses not only arguments for species being mutable but also the whole "Lamarckian system," including the supposed escalation, of animal and plant organization, from simple beginnings in infusorian organisms arising constantly in spontaneous generations. For we know that Darwin was probably making a fresh, close study of those chapters—in Lyell's fifth edition—in September 1838, one likely time for that recollection to be entered in the diary opened just before.[66]

Moving now to the second sentence, we can see that there are difficulties here, too. Late February and early March are indeed, as we shall see in due course, the most likely time for Darwin to have learned striking new conclusions about the Galapagos bird species (the mockingbirds and finches, especially) from John Gould, who was then making his reports on them to the Zoological Society, while conferring closely with Darwin.

But what was he then finding out about the South American fossils that he did not know before? Richard Owen had completed a preliminary identification of them in January and passed on his conclusions to Darwin before the end of that month. By the middle of February (the sixteenth in fact) Lyell was summarizing them in his presidential address to the Geological Society.[67] So, we must ask, why this memory of one month as peculiarly important in regard to both the fossils and the bird species?

And as for the last sentence, does it not smell somewhat of what we may call a naive empiricist myth of theory origin; and does it not smell no less of a property-staking claim? By all means, let us take it seriously; but warily, too. Darwin is well known as someone who has left us several brief and cryptic reminiscences as to how his ideas on various subjects originated. And most of them have turned out to be mistaken, not to say misleading, when we have been able to check them against independent evidence. To be sure, this particular reminiscence was apparently made within months or years rather than decades of the deeds. But is it not likely, at least, that its main purpose is to convince any reader, above all Darwin himself, that his theoretical views were strictly derived from facts, hard facts about his fossils and his birds, and not from theories, especially not from anyone else's theories about their facts?

But, even apart from such exegetical possibilities, there are ambiguities aplenty in the relationship asserted between what is supposed to have happened in March and the opening of Notebook B in July. Is the suggestion that some March findings then prompted a decision, implemented in July, to open such a notebook? Or did they actually prompt the views—and if so which ones—explored in that notebook? Did they elicit commitment for the first time to the mutability of species? Or did they rather suggest how it is that transmutation is caused? Or that there is much more evidence for transmutation, or for some thesis about transmutation, than had hitherto been thought? The possibilities are all too easily multiplied; from their multiplication alone, we can see that there is no way to get any clear indication, from these celebrated sentences, as to how the opening of the July notebook is related to whatever it was Darwin later remembered being struck by in March.

We need, then, a new place to start, in our efforts to clarify the relationship between the opening of Notebook B in July 1837 and the initial rationale for Darwin's first favoring of transmutation, whenever that occurred. For reasons that will emerge in due course, we do well to begin by looking at the way Notebook B itself opens, before considering its place in the whole succession of documents Darwin wrote in the first ten months after returning home in October 1836.

Several close students of Notebook B have lately appreciated that its opening entries (up to B 27 or 28 as far as I can judge from looking at the manuscript) form a continuous and coherent composition and set out a single comprehensive structure of argumentation.[68] These entries even appear (obvious later insertions apart) to have been made at a single sitting. So the most likely conclusion is that they record a first putting down on paper of several elaborate and connected trains of thought worked out in advance. And that conclusion fits well with the fact that up until June Darwin had been very busy preparing the text of what would appear, in 1839, as his *Journal of Researches.* This text was complete by the end of June and before the end of the summer he was correcting the proofs. So, July was a moment when it made sense to open that theoretical notebook and to do so with a burst of writing that comprehended all the main conjectures already being seriously entertained.

As his heading, across the first page, Darwin wrote *"Zoonomia."* It was, of course, the title of his grandfather's book, and that is soon being cited on a fundamental issue.[69] But it is clear that Darwin wants this heading as his own, not because he is writing out a response to his grandfather's book; but because he is inquiring into its subject, "the laws of life," and doing so, moreover, according to his own understanding and not his grandfather's of what such an inquiry into such a subject includes. The entries from B 1–27 (more precisely the first 160 lines in Barrett's print-out edition) may then be called Darwin's "Zoonomical Sketch" of 1837.[70]

There are several ways to analyze the structure of that sketch and we will dwell on one at length later. Here, however, we do best to consider it as comprising two parts (B 1–15 and B 18–27) linked by a transitional section (B 16–17). The first part can then be seen to take us all the way from sexual (as contrasted with asexual) generation, as allowing for adaptive variation from one generation to the next, to divergences as wide as those between the Australasian and other mammals; while the second makes a new start (after the transition), and takes us all the way from the simplest "monads," produced by spontaneous generation, to the higher organisms eventually arising in the progressive escalation and proliferation of organization in a whole "tree of life" or, better, a whole "coral of life." For the first starts with individual organisms reproducing sexually (rather than asexually) and then argues that—with enough eons of time and so changing circumstances—the adaptive variation, which we see daily due to sexual generation, can eventually lead to faunal and floral differences as large as any we find in the world; while the second starts from a monadic infusorian beginning of life and considers all the successive species, arising in the subsequent organizational escalation, as analogous to the successive propagating and dying of individuals making up a growing coral polyp colony or a growing tree interpreted as a bud colony.

Now, what emerges when one surveys the whole of this July 1837 Sketch, comparing it with the *Journal of Researches* text, is that the *Beagle* voyage findings were then taken by Darwin to bear, in many and various ways, on a multitude of distinguishable issues. Those findings themselves we may consider, in no particular order, as forming four clusters as follows:

1. The Galapagos fauna and flora. These bear on four issues:

*a.* Unlike those on the mainland, the birds there have no instinctive fear of man. The two lots of birds therefore show that new instincts can arise very gradually as heritable habits, and that (contrary to Lyell's view in criticizing Lamarck) such new instincts are not merely modifications of existing ones (B 4; cf. *JR,* chap. 19).

*b.* The Galapagos tortoises and mockingbirds, differing as they do from one island to another, support the view that a few individuals, perhaps a pair only, of a species would eventually give rise to a new species, following their migration and separation on a fresh island; as is also indicated by foxes and by hares elsewhere (B 7).

*c.* The Galapagos finches, as sorted into a dozen or so species by Gould, show a gradation of structure, especially in their beaks, which pass from a thick seed-eating beak at one extreme to a very slender insect-eating beak at the other. However, because every species is not on every island, the gradational series is not seen as long as one stays at a single spot. The finches thus support the generalization of Lamarck, as reported by Lyell, that character gaps between species are eventually filled in as collections over a whole area are made more complete (B 9 and *JR* 475).

*d.* The fact that many of the species peculiar to the Galapagos are of genera also represented, often exclusively, on the South American mainland instantiates a generalization confirmed by Juan Fernandez, an island that is much closer to the mainland; namely that the number of distinct species will depend on the distance from the mainland. If this is small, continued migrations of new settlers will prevent the descendants of earlier ones from diverging; while if it is large they will have time to become altered sufficiently not to breed with any later settlers (B 10 11).

2. The two *Rhea* (ostrich) species living in southeastern regions of South America, bear on four issues:

*a.* As with several other cases in that region, we find one species replaced by another, very similar species as we travel from the wetter northern to the dryer southern regions. This replacement with climate change indicates that a species can vary enough in altered circumstances to give rise to new species (B 7).

*b.* In the ostrich case, the two species "inosculate"—meaning, here, simply that their ranges overlap. So one must suppose that the younger southern species living on the newer land arose within the territory of the other, and so in a sudden, not a gradual, gradational, change; for such a divergence would be counteracted by continued crossing (B 8).

*c.* The ostriches, like the several extinct species of South American *Megatheria* (sloths) exemplify the general fact that a genus is often found exclusively in one continent, a finding explicable if all its species are descended from a single parental species. Such cases thus provide an answer to one of De Candolle's arguments as retailed by Lyell. Countering the proposal that new species arise from the hybridization of old ones, De Candolle had urged that often a species lives a long way and across barriers from any that are sufficiently like it to be plausible parents of it. Accordingly Darwin answers, in effect, that he does not need two parents for each species, only one for several, and that there are many cases where several congeneric species live in the same region (B 12–13; cf. RN 153).

*d.* As we travel in space today, we find the two ostriches succeed one another. If in space today, so perhaps in time of old. They indicate strongly, therefore, the possibility of new species arising from earlier species (B 17; cf. RN 130).

3. The South American fossils bear on six issues:

*a.* They offer cases of genera being exclusively found in one region at some one time (see 2*c*, above).

*b.* They instantiate strikingly the law of regional succession of peculiar genera, since the extinct species are often of genera now exclusively represented by extant species in the same region (B 14).

*c.* They provide the decisive evidence for Brocchian species senescence (to use Kohn's apt term) as the best explanation for individual species extinctions (B 21–22; cf. RN 128 and 133).

*d.* They confirm the generalization, insisted upon by Lyell, that mammal species do not last as long as mollusk species. This generalization can be explained by supposing that any monad and all its issue has a limited total lifetime; so that those monad issues that have progressed furthest, and so progressed to mammalhood, have changed in species at the highest rate (B 22–23 and RN, inside cover).

*e.* Some lines of descent have ended without further progeny, the *Megatheria* lines for example in South America; while others, such as some of the armadillo lines, have persisted thanks to species transmutations (B 20).

*f.* Finally, one sees that some branches must end without splitting so that others can split without ending; otherwise, in a world fully stocked with living beings, there would be no room for the new species. The South American fossils exemplify this thesis too (B 20–21).

4. The Australasian mammals bear on four issues:

*a.* There are many antelope species in South Africa, but in similar conditions in Australia there are marsupial species instead. In neither case, then, is the common structure characteristic of that group of species necessitated as an adaptation to those conditions. But it could well have been necessary for the single species ancestral to each group (B 14–15).

*b.* The countries longest separated have the greatest differences in their indigenous species—the Asian placentals and Australian marsupials being the most striking case (B 15–16).

*c.* If separated for an immense age, then the indigenous fauna in each country (e.g., Asian and Australian mammals) will have diversified, so as to be represented by distinct congeneric species at different places within each country.

*d.* Such an explanation for the distinctive Australian mammal fauna presupposes that there were once no mammals, or at least no placentals, on earth at all, and that the Australian marsupial mammals are descended from a different stock than those in the rest of the world (B 15).

We are now much better prepared to understand what is going on in and between the lines entered by Darwin, a few months before he opened Notebook B, in seven pages, now scarcely less famous, in his Red Notebook (RN 127–33). These pages have benefited recently from close exegesis (especially by Herbert and by Kohn). However, the full range of their concerns has not been explored sufficiently. No apology is made here, then, for doing what has not been done before: that is, going through entry by entry to see how many of the July issues were in play at the earlier time. Obviously, in such a quest precautions must be taken against anachronistic interpretation. But, as we shall see, the price of not running that risk and taking such precautions is to misunderstand the texts themselves.

To look ahead a moment, we need to note a clear trend in Darwinian scholarship as it concerns these months of late 1836 and early 1837. In an older generation, it was customary to comb through the published books, papers, and correspondence of the *Beagle* years and to pronounce, on the slightest evidence, that Darwin was showing himself to be pondering the transmutation of species even as he dug up his fossils or collected his finches. This custom has now very properly been replaced by a much more circumspect approach; so that only hard, direct, textual traces of transmutationist thoughts are taken seriously into account. This new circumspection marks a great advance, obviously, not least because it has been exercised upon a much wider range of documents than used to be studied. Inevitably, however, the pendulum has swung too far. The current consensus places Darwin's first serious inclining to transmutation no earlier than the winter of 1836–37, and perhaps as late as March. But in fact the interpretation of the evidence that this consensus requires is not really plausible. In projecting its own caution onto Darwin, the new scholarship finds only tentative transmutationist commitments behind his cryptic notebook entries at this time; but, ironically perhaps, it has thereby brought out how the textual data themselves must now lead us to start the pendulum swinging back the other way, although only a matter of months rather than years.[71]

Before proceeding to the Red Notebook itself, we should recall where it belongs in the whole succession of private and public documents that

Darwin was producing in the months between landing back in England (in October 1836) and opening Notebook B the following July. Those months fall into three periods. In the first two months, Darwin was traveling a great deal, mostly between Shropshire (the family home), London, and Cambridge; and he was trying to settle the disposition of his *Beagle* voyage material. He was, consequently, getting rather little writing done. Then from early December to early March he was mostly in Cambridge and, despite an active social life, making a start on his writing, including his *Journal of Researches.* From early March he was in London, where the writing went much more quickly, with the *Journal* having a much larger share of his time.[72]

The composing of the *Journal* consisted mainly of integrating scientific samplings, from his zoological and geological diaries and from his ornithological notes, with extracts from his personal diary. We can only guess how he proceeded, but it seems most likely that he prepared his draft chapters in the order they eventually appeared in, while allowing himself on occasion to go back to chapters already written, to put in additions and corrections. We may, therefore, ask when any chapter may have been written initially. Reading Darwin's progress reports in letters to Henslow (March 28, May 18, and May 28) and allowing for the much quicker progress made after leaving Cambridge, it seems reasonable to guess that he had no more than six to eight of the eventual twenty-three chapters in draft when he moved to London in early March.[73] If so, then the first chapter to be extensively concerned with the South American fossils (chap. 5) was likely written in February, and the second to be so concerned (chap. 9) in March. It could be, therefore, that March seemed an especially important month, in retrospect at least, because it was then that his mind was concentrated both on the law of the regional succession of peculiar types (so prominent in *JR,* chap. 9) and on the Galapagos mockingbirds and finches that Gould was reporting on to Darwin and others.

Such a conjectured coincidence is certainly consistent with other documentary sources: especially the presentations, in March and in May, that Darwin made to the Geological Society and to the Zoological Society, which concerned the finches, the *Rhea* species, and the fossils; it is also consistent with the relevant parts of the Red Notebook.[74]

As Herbert has shown, the 178 or so pages of RN entries comprise two sequences—those (through p. 112) mostly made in pencil during the last months of the voyage, from about May to late September 1836; and those (pp. 113–78) made mostly in ink, after his return, in the months ending, probably, with June 1837. To Herbert's arguments for dating the second sequence of entries, there is only one additional observation to be made here. At page 113 itself, there is a reference to a "camel." And it is clear that this refers to the Port St. Julian fossil, as now correctly identified by Owen (some time in January) as an extinct camelid species, allied therefore to the extant llama or guanaco species that is living in the same region. To be sure, the camel species now living in Asia was mentioned, as early as the February 1835 note, as a species of comparable habits and diet; but it is clear when one compares RN 113 with *JR*, chapter 5 that the reference here is to Owen's identification of the Port St. Julian bones themselves. The ink entries run, then, from January to June 1837.[75]

This point about RN 113 thus confirms Herbert's observation that the RN entries from page 113 on relate to the writing of the *Journal*. And it seems reasonable to associate page 113 in time as in topic with *JR*, chapter 5. Accordingly, the topics taken up at RN, 127–33 would be associated in time and topic with the writing of *JR*, chapter 9. Herbert's suggestion of March for those entries is thus reinforced. It is also reinforced by seeing that RN, pages 125 to 160 are consistently occupied with topics relating to the South American mainland, as those topics are about to be discussed in what will be *JR* chapters 9 through 18. (This is especially true regarding the Cordillera of the Andes, as discussed in chapter 17.)

This relation between this stretch of RN and those chapters of *JR* alerts us, therefore, to the possibility that, even in the privacy of RN, Darwin is often constrained by his perception of what would be appropriate in the *Journal* as a publication. Equally, the extreme brevity of most of the entries is no indication that their thoughts are fleeting or peripheral ones. As promissory memoranda they represent not so much the full extent of Darwin's reflections at the time, as his decisions as to how to work various topics in with one another when the time comes to do them justice in the *Journal*.

This point is especially borne out when we see what is apparent from pages 127–33 themselves (or from a photographic reproduction such as Herbert provides for pp. 127–30) but not always apparent from the transcriptions published: namely, that the entries there fall into ten often very brief clusters, each marked by Darwin by a horizontal line if it ends before the end of a page.

Seven of these clusters concern us here. The second and fourth (both on p. 128) do not; they concern the contrast between the climate of the northern and southern hemispheres (see *JR*, chap. 13) and the presence of life in hostile habitats high in the Andes (see a footnote added, later presumably, at *JR* 77 and also chap. 13); while the eighth (the only one on p. 131) discusses the thickness of the earth's crust as shown by the Andes. Going now to the first such cluster (RN 127) it reads:

Speculate on neutral ground for 2 ostriches; bigger one encroaches on smaller.—change not progressife [sic]: produced at one blow. if one species altered: Mem: my idea of Volc islands. elevated. then peculiar plants created. if for such mere points; then any mountains. one is falsely less surprised at new creation for large.—Australias = if for volc. isld then for any spot of land. = Yet new creation affected by Halo of neighbouring continent: = as if any creation taking place [these last two words inserted] over certain area must have peculiar character:

Darwin is obviously doing two things here and relating them to one another. He is resolving to speculate on the implication of the overlap in territory of the two ostrich (*Rhea*) species on the mainland: namely, that if one has arisen from the other then there has been a sudden transmutation. And he is recalling an associated idea regarding species origins on volcanic islands.

So, in the prospective *Rhea* speculation there is, in the "if," no implicit tentativeness about transmutation itself (as Kohn has suggested).[76] For as Notebook B and RN 153 indicate, the implied contrast is between two transmutation possibilities: either one of the two species now extant has given rise to the other or both have arisen in the alteration of a common parental species. And the reasoning is that, if the former had happened, then the alteration must have been sudden, not gradual; witness the territorial overlap.

That this is the implied contrast is confirmed, when we see Darwin going on to a further related contrast: species formation by transmutation on islands as contrasted with mainland areas. On new land continuous with existing land, migration over land and saltation of territorially overlapping species may have been usual. But when the new land is a volcanic island in the ocean, animal migration or plant transport of a few segregated settlers from the mainland has preceded the origin of the new species. Accordingly, Darwin concentrates on the fact that even tiny points, volcanic oceanic islands, often have their own peculiar species. Now, these species, following Lyell's phrasing, he describes as "created" for these islands. But this very source of the phrasing shows that his argument does not take them to be separate creations, species originating independently of all others. For, were they so taken, there would be no explanation (only a jokey, absurd pseudo-explanation invoking a "halo") of their similarity to species now living on the nearest continent. For this similarity presents the same explanatory challenge that is presented by any area containing peculiar genera, families, or orders—an area of continuous land such as Australia or an area of land and sea as that which includes, for example, western South America and the Galapagos Islands.

Although the phrasing is from the *Principles,* Lyell's account of the issue is being deliberately rejected by Darwin as incomplete. Lyell had recounted how St. Helena was, when first discovered in 1501, "covered with a forest of trees and shrubs, all of species peculiar . . . with one or two exceptions, and which seem to have been expressly created for this remote and insulated spot." He had also noted that, "although it has justly excited great attention," the "exclusive occupation" of Australia by various tribes of marsupials is strictly in "accordance with the general law of the distribution of species." For elsewhere around the world "we find peculiarities of form, structure, and habit, in birds, reptiles, insects or plants, confined entirely to one hemisphere, or one continent, and sometimes to much narrower limits."[77]

To this argument, Darwin is replying that, yes, a big mass of land such as Australia is, biogeographically, no more surprising than a point such as a volcanic island that has its own species. But there still remains the challenge raised by both: explaining the close similarity, and so the peculiar generic or familial character, of many of the species originating throughout the whole region in each case.

That this is Darwin's argument here is confirmed when we notice that a sequence of six pages (at B 90–103) written in autumn of 1837 goes over almost exactly the same argumentative ground, as do this RN 127–28 entry and the related entry at RN 128–29; and when we notice that those Notebook B pages likewise retain the talk of creation and employ ironic phrasing reminiscent of the earlier halo talk: "The question if creative power acted at Galapagos, it so acted that birds with plumage and tone of idea purely American, North and South,—so permanent a breath cannot reside in space before island existed.—Such an influence must exist in such spots. We know birds do arrive and seeds" (B 98). In other words, here, as early in the spring, two ways of attempting to explain the similarity of the island and the neighboring continental species are being contrasted: a mysterious halo or breath, versus migration with subsequent transmutation. The spring entry is thus sketching this line of argument and, moreover, recalling it as one already formulated even earlier.

This interpretation is borne out when we pass (skipping the second cluster, as explained just now) to the third, which reads: "Great contrast of two sides of Cordillera, where climate similar—I do not know botanically = but picturesquely = Both N and S great contrast from nature of climate =." For what is being drawn here (compare *JR* 499, the *Beagle* diary for March 23, 1835, with the Sketch of 1842 and Essay of 1844) are two complementary contrasts. First, on either side of the Andes, there is a striking contrast, as one travels from west to east, in the character of the animal species (and perhaps the plant species, too), even though the climate is similar. Second, on both sides of the Andes, if one travels north and south there is seen a marked contrast in the species and marked changes in the climate. So, putting these two contrasts together, the whole entry is insisting that the character of the species (and so their supraspecific groups) in two places may contrast both when there is not and when there is a contrast in climatic conditions. So, this thesis obviously complements the one implicit in the discussion of oceanic islands and neighbouring continents; for there one finds a similarity of distinct species and a dissimilarity of conditions (most strikingly in the Galapagos which are so unlike the nearest South American coastal lands in soil, climate, and so on). Darwin has, therefore, got three permutations now deliberately in play: similar indigenous species with dissimilar conditions; dissimilar indigenous species with similar conditions; and dissimilar indigenous species with dissimilar conditions.

So, it is plain that the principal issue which Darwin sees to be at stake here is nothing less than the quite general one of how far adaptation to conditions, especially climate, can explain the character of the species originating in any area. And it is surely evident that Darwin already holds, and has done so for some time, that (contra Lyell) it cannot explain it adequately at all.

The remaining entry clusters make that exegesis even more secure. Skipping the fourth, as explained, the next one, running from page 128 to 129, records a resolution to establish those facts about the extent or limits of territory and migration that are relevant to both the mainland and the oceanic island cases already discussed: "Go steadily through all the limits of birds and animals in S. America. Zorilla [a skunk species]," he says, continuing on the next page: "wide limits of Waders: Ascension. Keeling: at sea so commonly seen. at long distances; generally first arrives:—" The point here is, therefore (see *JR* 543, for example) that mammals do not migrate so readily as birds do across barriers on land masses, such as mountain ranges (e.g., the Andes); and that certain birds, waders especially, are the ones that range farthest overseas and the ones that are present but (unlike finches and mockingbirds) without peculiar species, on oceanic islands. And it is this point that is complemented in the next entry: "New Zealand rats are offering in the history of rats, in the antipodes a parallel case." As with islands like St. Helena in the Northern Hemisphere, lacking indigenous mammal species, the odd rat or mouse species excepted, so with New Zealand, which had but one indigenous placental mammal species, a rat; thereby showing that the insular representation of a supraspecific group depends on the migratory powers of that kind of animal and not on the conditions in the island, whether small, like St. Helena, or large, like New Zealand (cf. RN 62 and 79).

This theme—of the explanatory inadequacy, in historical biogeography, of adaptation to conditions—can now be seen to dominate and connect the remaining entries on these pages (excepting the one on p. 131 as explained). As a species, the "extinct Llama," of Port St. Julian, owed its death, Darwin notes, not to a change of circumstances, for there was none. He is, then, "tempted" to believe generally that animal species are "created for a definite time: not extinguished by change of circumstances." Further, as the two ostrich species succeed and overlap one another in space today, so we must suppose that the extinct and extant llama species did so in their day. Hence: "not *gradual* change or degeneration. from circumstances: if one species does change into another it must be per saltum—or species may perish." The older species could not go on living, as such, beyond its limited lifetime, but by changing into another it could have a new lease of life as a new species in the same conditions. In general, then (and Darwin now cites other cases matching the *Rhea* one) the representation of a genus by two or more species in a single region indicates transmutation; their territorial overlap within that region indicates, as Darwin argues, sudden rather than gradual transmutation (RN, 130).[78]

Hence, therefore, the analogy between the way adaptation to conditions shows its explanatory inadequacy in the face of certain extinction facts, and the way it does so in the face of certain species origin facts. Having elaborated (p. 132) his arguments about all propagation, sexual or otherwise, reducing to a division of an individual, and having thereby supported his Brocchian extinction theory (see the discussion in section 5 of this essay), Darwin can now make the analogy explicit (at p. 133): "Dogs. Cats. Horses, Cattle. Goat. Asses. have all run wild and bred. no doubt with perfect success.—showing non Creation does not bear upon solely adaptation of animals." Once again, Darwin is turning facts reported in one place in Lyell's *Principles* (and since confirmed by his own observations) against theories expanded elsewhere in the *Principles.* The almost invariable absence of indigenous mammal quadruped species on oceanic tropical islands, such as the Galapagos, provided Lyell with a very important analogy to the European Carboniferous fauna as he interpreted it. For he argued that these facts about oceanic islands in tropical regions, today, are presumably instances of a general law that such sites are always unsuited for mammal quadruped life, so that no such species would originate on them. And yet in discussing the migratory powers of mammal quadrupeds, he emphasized how widely domesticated mammal species, such as dogs and cats, have been able to spread, even colonizing many islands around the world, those in tropical climes not excepted.[79] These and any other such facts about these domesticated mammal settlers run wild were, therefore, welcomed by Darwin as evidence for the case he was marshaling in the spring of 1837. For they were striking instances of how the absence of indigenous mammalian quadruped species (their "non Creation") on most remote islands—their consistent failure, so to speak, to have originated in such places—cannot be explained as Lyell would explain it: as due to a lack of adaptation of that kind of organization to that sort of site. Darwin accordingly closes his entry by making explicit the analogy with the inadequacy of the explanation for extinction, failure to survive, that Lyell had offered: "extinction in same manner may not depend [solely upon adaptation].—There is no more wonder in extinction of species than of individual" (RN 133).

So, taking these pages in their context, what should we conclude is new here, and what not, in Darwin's thinking about species origins? Surely, that he may well have been working out, at this time, as he had not before, the contrasting modes of species formation for mainland and for island situations, that is, respectively, sudden, sympatric formation of one species from another (the Patagonian *Rhea* paradigm) and gradual, allopatric species formation by divergence following isolation (the Galapagos *Orpheus*—mockingbird—paradigm). It is quite implausible, however, to see him as tentatively exploring transmutation itself for the first time. For it is surely plain that he has already, and for some time, been organizing his thinking around a very general opposition: that between, on the one hand, Lyell's explanations of new species characters as determined by adaptation to conditions, and, on the other, his own alternative explanation of such characters as dependent on migration and transmutation.

To accept this view of Darwin, in the spring of 1837, is still to leave open several possibilities. Had his initial rejection of Lyell's explanatory appeal to adaptation, in explaining geographical distribution of extant and extinct species, been accompanied from the start by a simultaneous favoring of transmutation as a preferable alternative? Or had some other alternative been contemplated at first, itself later rejected in favor of transmutation? Or had there been a period when Lyell's theory was rejected but none favored in its place? There is, needless to say, no direct documentary evidence allowing us to decide among these possibilities without hesitation. However, we shall see in due course that the indirect evidence all makes the first of the three the most likely.

Two considerations might have inclined us to think it likely independently of any further, circumstantial clues. For first, over the problem of extinction, Darwin had gone at once, on rejecting Lyell's own theory, to favor a theory respectfully rejected by Lyell: Brocchi's. We might expect, then, that on rejecting Lyell's theory of species origins he would do likewise; and immediately consider whether the only other theory given full and respectful discussion by Lyell—transmutation—might hold up where Lyell's failed. Second, transmutation, in so far as it allowed for the possibility of a common descent for similar but distinct species, plainly provided for an extension of the kind of genealogical explanation for biogeographical data that Lyell took over from De Candolle and others. They, as we saw, had explicitly considered the problem of explaining why individual species are often missing from areas that suit them excellently once they are transported there; and these authors had, of course, explained such absences as due to all the members of any one species having descended from a single stock, at a single spot, with the descendants since confined by barriers to some only of the regions where they could in fact flourish. Thus was common ancestry used to explain what peculiar adaptation could not. Likewise, then, consider a close student of the *Principles* with Darwin's reasons for thinking adaptation inadequate to explain any cases of the common confinement of all the species of some genus to one side of a barrier, when the territory beyond that barrier is exactly similar in climate and other conditions and so equally suitable for these species. We might expect this young Lyellian to contemplate a common ancestry for congeneric species; as explaining what cannot be explained by supposing them all peculiarly adapted to conditions peculiar to the region wherein they are all now confined and wherein they all originated.

Finally, two further remarks about the documentary probabilities. Virtually only one known extant document, the February 1835 note, stands between us and the conclusion that Darwin's Brocchian extinction theory was first adopted by him in the early months of 1837. His long-standing commitment to that theory over the previous two years after February 1835 has apparently left no other known extant trace. By parity, then, perhaps the suggestion that Darwin was committed to transmutation well before the spring of 1837 should not be thought conclusively refuted merely by the lack, almost total as it is, of explicit traces of such a commitment in the extant documents as now known. Recall, too, that Darwin's discussion of biogeographical generalizations in the Essay of 1844 often keeps the commitment to transmutation strictly implicit, sometimes for pages on end; while even the phraseology of "creation" is occasionally retained and not always with obvious ironic intent. There are, then, several reasons why we should not be surprised at finding no sustained, overt expression of Darwin's transmutationist dissent from Lyell in the documents surviving from 1836.

As far as positive expectations go, we have seen a single conclusion suggested by the relevant Notebook B and Red Notebook texts from the summer and spring of 1837: namely, that the *Beagle* voyage findings were then taken to bear in many ways on many issues concerning the laws of the animate part of the terrestrial system; but that the particular issue, of whether or not species originate by transmutation, was held to be raised most directly by one problem: how far adaptation to the conditions there can explain the characters (and so the supraspecific group representation) of the new species originating on new land (insular or continental) arising by elevation from submarine depths; and, conversely, how far adaptation to conditions must be supplemented with migration and descent from species already living on nearby older land. This conclusion, from the 1837 texts, can, needless to say, only raise a tentative presumption that this problem had originally provided the initial rationale for Darwin's earliest disagreement with Lyell on this issue. As it happens, however, there is sufficient independent circumstantial evidence from the previous two years to put that presumption beyond most reasonable doubts. It is to this evidence that we must now turn.

## 8. Darwin's Movements away from His Early 1835 Position

Before considering the evidence itself, we need to outline briefly the overall shape of Darwin's experiences and reflections in the last year and a half of the voyage (ending in October 1836). We left him, so to speak, some pages back, on the west coast of South America in February 1835. This was three years after he had first reached that continent, and some six months before he would leave it for the Galapagos Islands where he would enter the last year of his travels.

Even these bare chronological details alert us to decisive asymmetries and disproportions in his five years study abroad. Not only was southern South America obviously the dominant territory; it was the only continental region that he explored with any thoroughness at all. To be sure he would have a week or two in Australia and also in South Africa; but the geology and zoology of those regions he would never make his own, as he had those of southern South America. No, once South America was behind him, it was the islands, not the continents of the Pacific, Indian, and Atlantic oceans, that he would master in observation and theory.

The relation between the earlier continental and the later insular experience is, therefore, of prime significance. And we do not have to guess at Darwin's own view of its significance at the time. South America was, even as he stood upon it, jolted by its quakings, a mass of land in elevation, one whose highest mountains had once been a string of volcanic islands thrust above the ocean waves by successive upheavals from below. For a protégé of Lyell, elevation here implied balancing subsidence elsewhere. And, sure enough, even before leaving South America and when he had yet to explore one himself, Darwin was following Lyell in speculating on the connection between coral island formation and subsidence of the earth's crust. A cliché has it that a hungry animal divides the world into edible and inedible objects. Darwin, hungering for support for his Lyellian speculations about these risings and sinkings, divided the islands of three oceans and eighteen months' traveling into those whose elevations were marked by visible volcanic craters and those whose subsidences were commemorated by hidden coralline monuments.[80]

For his thinking about the origins of species, what was most instructive was the contrast between the incipient faunas and floras of the new land starting as volcanic islands, and the remnant faunas and floras of the shrinking land sinking beneath coral islands. In the last months of the voyage, his mind was concentrated upon this contrast, especially by one exemplar of each type of island: Keeling, a group of coral islands in the Indian Ocean, visited in April 1836, and St. Helena, a single volcanic island, visited fourteen weeks later, in July.[81]

The long months sailing across the last two oceans, with the end of the voyage imminent, were naturally used to put in order records and specimens from the whole five years; and not only physically, but theoretically too. So they were a time to ponder any theoretical speculations that might link the sites being visited with those that had yielded, long before, the records and specimens now being considered as material for professional specialists to study following the return to London. Thus, then, we are alerted to the significance of some possible coincidences. Darwin would later (in the spring of 1837, in the Red Notebook, as we saw in our last section) remind himself of an idea he had had about peculiar species originating on islands newly elevated by volcanic eruptions, and he would do so in phrases echoing to a word those used by Lyell of St. Helena. Now, the time (mid-1836) when Darwin was himself in the Atlantic and visiting this and other comparable islands, such as Ascension, is as likely as any other to have been the time when he wrote up, in his *Ornithological Notes,* his first full account of the Galapagos Islands; this is an account, moreover, which relates their whole fauna and flora, not merely their birds, to their physical history and geography as distant, oceanic, volcanic islands. As we shall see, the account presupposes throughout that the indigenous species have originated by transmutation, and it does so in a way that indicates that Darwin had already embraced that presupposition some time before; although exactly when we cannot tell from that document or any other.

It looks, then, as though there could be another parallel between Darwin's break with Lyell on this issue and his earlier one over extinction. Darwin had left Lyell for Brocchi on reflecting about the Pampas and Patagonian fossils collected a year, and sometimes more, beforehand. Likewise, by the time he was exploring the volcanic islands of the Atlantic, and apparently pondering anew those in the Pacific, he had been a Lyellian student of oceanic island faunas and floras for nearly a year, starting with the Galapagos (visited in September and October 1835).

Moreover, if this narrative is accepted as a fair conjecture, then we have a resolution of some seeming paradoxes. In the spring of 1837 the Galapagos fauna and flora seem to have been very important to Darwin for new reasons, and yet he had plainly been attending to them very carefully well before that, in 1836, even if he had been hardly so excited by them still earlier when actually on the islands in 1835. But all three moments make sense on the account here proposed. When first encountered, that fauna and flora could not be viewed as an instance of any general theory of island historical biogeography; but would be primarily of interest as geologically recent and geographically remote additions to the South American fauna and flora. A year later they had become thoroughly assimilated, along with St. Helena and others, as a striking instance of a comprehensive speculation concerning new species for new land, including merest points of land arising in oceanic, volcanic eruptions. Then after a further year they could be additionally assimilated into further generalizations (as St. Helena obviously could not) specifically about island clusters, archipelagoes. So, in sum, to their initial significance, primarily as South American, there could be added successively their later general significance as volcanic and, then, still later, as archipelagic. But the Galapagos were sometimes cart, sometimes horse, and sometimes neither. Initially, they were associated with no new thoughts about species origins. Then, apparently in 1836, a new general thought (transmutation) brought a changed view of the Galapagos; while, the following year, 1837, it was new conclusions about the Galapagos species that brought a new view about species transmutations (the efficacy of isolation on islands in initiating divergence).

Such schematic and speculative suggestions about Darwin's thinking in these two years, 1835 and 1836, would be merely fanciful, of course, were they not constrained by reasonably reliable documentary data. For our purposes here the principal documents are three which can be dated pretty confidently to February 1835, and one, the *Ornithological Notes,* which, as already mentioned, is probably assignable to the middle of 1836. In addition, various clues can be gathered from texts in the months falling between these two dates.

From February 1835 we have not only the note on the Port St. Julian bones and extinction and the "Reflections" on Buckland's appendix to Beechey (discussed earlier in this paper), we also have an associated set of ten pages headed by Darwin "Reflections on reading my Geological notes." These have already been tentatively placed in 1835 by Herbert and by Gautrey.[82] Several features of them make it highly probable that they are from the same month (February) of that year. These pages were apparently kept by Darwin himself with these others. Their title matches that used for the pages on Buckland. In their pen, ink, and handwriting they seem similar to the others; and their prose is similarly scrappy, telegrammatic, full of deletions, and so contrasting with the much more polished manuscript draft pieces on South American geology that evidently date from the early months of 1836.[83] Most conclusively of all, they refer several times to the Andes, but in tentative ways that never indicate firsthand acquaintance with Andean geology (see especially p. 5 recto); and, sure enough, Darwin's close examination of the Andes came in mid-March to late June 1835. Finally, they are, like the Port St. Julian note, prominently preoccupied with the question of the relative chronology of the mollusk and quadruped species. On the second page, moreover, it is conjectured that some extinct mollusks were "destroyed during the elevation of the superior Tosca"; and then this phrase is changed to read "themselves perishing during the deposition of the superior Tosca." They thus appear to record changes of mind about extinctions.

These reflections seem to be, indeed, the very first version of an exposition that is familiar from several printed version in later years. It starts by discussing the extent of the Pampas formation of the Buenos Aires region; and eventually moves, as Darwin himself had, of course, all the way south to Patagonia, while arguing (see p. 6 recto) that the land in these two regions should be seen as "one grand formation." Accordingly, Lyellian historical biogeographical issues are raised and responded to at two points: first in regard to the northern, and, second, the southern region.

Thus (at p. 7 recto) he sketches, with many corrections, an appropriate historical biogeography to go with his vast sequence of "repeated elevations." Initially, "rocks from seas too deep for life" were thrust up rapidly to a depth suitable for those extinct species, of oysters and so on, now found inland as fossils. Then further elevation produced the land whereon the "great quadrupeds lived," before their extinctions, while the former marine species "perished," to be replaced by those now found in the sea and scattered, as shells, on the land surface. Still later, further elevations have occurred, while the smaller extant quadrupeds have "roamed" the plains. And "doubtless" new beds are even now forming "beneath the ocean, ready when compelled to give their evidence in the open day light." On passing to the southern region, Darwin conjectures, likewise, a Lyellian historical biogeography: Patagonia has "risen from the waters at so late a period" that "it may be interesting to consider whence came its organised beings." His speculation starts by noting that "I have conjectured the absence of trees in the fertile Pampas and rich valleys of B. Oriental [two northern areas], to be owing to no creation having taken place subsequently to the formation of the superior Tosca bed." The implication is, then, that the Patagonian plains, being even more recent, must have been stocked with migrants from older land, because they will have had even less time for indigenous species to have originated upon them. Of the plants, he says, "I know nothing"; and a deleted phrase notes: "Insects are few probably." But "larger animals" than insects are "of easier knowledge." And so he makes a careful inventory of the land mammals. Confining himself to those south of the Plata River—because that area has "less immediate communication with Brazil" to the north—he notes that they nearly all are "characterized by a large Geographical range.—and therefore may easily have travelled far from their Northern original homes." And he gives some dozen and a half examples, starting with the "Mephites or Skunk," and including such species as the different armadillos (*Dasypus*), the peccaries, guanaco, and cavies. He ends by identifying two southern species as counterinstances to his generalization about wide geographical ranges. "The only two good exceptions. which I could not find out or see in South to occur further N. than 37°30′ is the Cavia Patagonica - and Dasypus pichy: (Azara says: Latitude between 35°–36°)."[84]

Taken together with the February 1835 note on the Port St. Julian bones, these conjectures show how completely Darwin had learned his lessons from Lyell on the origins of species. A region that is short of species both indigenous and otherwise is both recent and inaccessible; a region that is short only of indigenous species is recent and not inaccessible. Thus is the very recent but well-stocked Patagonia assimilated to Lyell's Sicilian precedent. Several months before, when at Valparaiso for the first time (in July and August 1834), Darwin had invoked the same explanation for the shortage of animals on the coastal land there. With plentiful bush vegetation he was "surprised to find that insects are far from common"; and indeed there was a similar "scarcity" of "higher orders" of animals: "quadrupeds" and "birds." But beds of shells of extant species, still retaining their colors, indicated very recent elevation. So it had seemed "not a very improbable conjecture that the want of animals may be owing to none having been created since this country was raised from the sea."[85]

Going back to the note on the Port St. Julian bones, we may recall how, in the older northern region in the east, the extinct and extant cavies of Bahia Blanca provided Darwin with a new instance, beyond those cited by Lyell himself, of the correlation over time of certain genera with certain districts. The same genus indicating the same conditions, this correlation supported Lyell's gradual succession of species, presumably by showing that extinctions do not come in batches as consequences of sudden changes in physical circumstances (see our discussion earlier in this paper). Talking of extinctions brought on by climate change, Lyell had said that "unless it pleased the Author of Nature that the planet should be uninhabited, new species would be substituted," to replace those extinguished by such causes. And Darwin had used the very same phrasing in affirming that continual losses from Brocchian extinctions would be providentially made good by continual species origins.[86]

Darwin's conscious commitment to Lyell's views on species origins is, then, complete and confident early in 1835. We may ask next, therefore, whether the rest of that year saw any shift from those convictions.

The ten months divide naturally into three periods—Andean, Galapagan, and Tahitian—prior to arrival (late in December) in New Zealand. There is no indication that the Andes prompted any reconsiderations. His geological conjectures there, as communicated to Henslow especially, concluded with the view that there were two mountain chains of entirely different ages. The more westerly was ancient, of a Secondary period; the more easterly was very recent because of the very same formation of the Pampas plains themselves. Both provided welcome vindication of Lyell: the westerly did so because its formation had the structure of the Tertiary but the age of the Secondary formations of Europe, thus confirming Lyell's view that "the crust of the world goes on changing in a Circle."[87] Originally raised up from submarine levels, it once showed above the ocean as no more than a chain of volcanic islands. The easterly confirmed Lyell, by indicating vigorous elevationary activity persisting undiminished into much more recent times.[88]

Now, Darwin records being struck, when there, by the contrast in the animals and vegetation on the eastern and on the western sides of the Andes, and, even more so, by the identity of the animals and plants high up in the eastern Andean slopes and plains with those familiar to him on the Atlantic coast, including Patagonia. But he does not seem to have asked at the time, how far this contrast and this identity could be explained by the geology of the whole region, nor how far the similarity in conditions on the two sides of the mountains was anomalous for a Lyellian historical biogeographer; even though he returned more than once to add new notes, especially about the birds, in his zoological diary.[89]

Nor do we find any signs of a changing mind in the notes Darwin made when at the Galapagos. Intriguing though they were, the islands conformed to several expectations. As promised, they contained active volcanoes, the first Darwin had ever seen, together with abundant signs (especially in their craters and in their coral life) of recent elevation. Again, Darwin had wondered in advance which "centre of creation" their inhabitants belonged to, and, sure enough, he was soon explicitly recognizing South America, especially in their "ornithology."[90] Finally, of course, Lyell had identified such equatorial archipelagoes, with their lack of mammals and abundant reptiles, as modern analogues of those he thought had dotted the northern seas at the time of the Carboniferous rocks. Just, then, as the Andes had shown Darwin a Secondary age formation of Tertiary character, so he could interpret the Galapagos as showing how formations of Secondary character could even now be accumulating in the Pacific.

On the tortoises, finches, and mockingbirds careful comments were recorded. It is said, Darwin noted, that slight variations of the tortoise shells are "constant according to the Island which they inhabit." Of the birds generally, he noted that by far the "preponderant number of individuals belongs to the Finches and Grosbeaks," although there seemed "much difficulty in ascertaining the species" among them. It was with mockingbirds that the most definite conclusions were reached. "This bird which is so closely allied to the Thenca of Chili (Callandra of B. Ayres) is singular from existing as varieties or distinct species in the different islands." More particularly, among the four individuals collected as specimens, there "will be found to be 2 or 3 varieties. Each variety is constant in its own island . . . a parallel fact to the one . . . about the Tortoises."[91]

In general, then, there appears no trace of any attempt to explain, while there, why the islands collectively or individually should have the species, and the supraspecific groups, that they do have and not those that they lack, whether indigenous or otherwise.

Within a year, however, in the *Ornithological Notes,* Darwin would be giving that challenge very careful attention. So we have to ask what may have happened in the intervening months that determined him to respond to it as he did. One region, and the reflections it prompted at the time Darwin was there, stands out as a very suggestive candidate: Australasia.

Between the Galapagos and New Zealand, Tahiti had stirred Darwin and Fitzroy to write in defence of the missionaries; and the island had interested Darwin geologically, but it was not one that yielded much in the way of observation or theory in botany or zoology. By contrast, within a few days of arriving in New Zealand, Darwin was noticing what was not there: "In the woods I saw very few birds"; and as for "animals" (that is mammals, as distinct from birds), it was remarkable that such a "large" island, with such "varied stations" should have not one "indigenous animal," excepting a small rat. Now, this last fact was no new finding, as Darwin knew. The remarkable conclusion had been established in the previous generation: one or two possible exceptions apart, Australia had no placental mammals indigenous to it, only marsupials; and, even more strikingly perhaps, New Zealand, again one or two exceptions apart, had none of either kind, placental or marsupial. Darwin would have been very much aware of New Zealand's lack, if only because Lyell had cited that country as a splendid modern southern analogue of the lack of mammals on his Secondary age northern archipelagoes. But he also notes that the indigenous rat was being "annihilated" by the European species introduced only two years before, and he records further examples, among plants, of European species overrunning this territory. Three weeks later he was in Australia and carefully recording other instances of European plants and animals doing exactly what he could see the European men were doing: namely, taking over the area by beating the natives on their home ground. It is records of those observations that provide the prelude to a famous passage where Darwin explains how he lay on a sunny bank and reflected "on the strange character of the animals of this country as compared to the rest of the World."[92]

The passage has often been misunderstood. It features two imaginary theorists, a "Disbeliever" (theologically) and "a Geologist"; and it takes off from a passage in Lyell, who was, in turn, responding to one in Paley.

After quoting Paley on the unity of creation throughout all space, Lyell noted that the geologist can pass in imagination rapidly from one period in time to another very remote from it. Even more than the traveler in space, the geologist is liable to infer wrongly that he has passed from one natural order to another, from the laws of one Creator to those of a second.[93]

Darwin, on his sunny bank, had been musing that anyone relying on his reason alone might see the contrast between the Australian fauna and the rest of the world's as evidence that "two distinct Creators have been at work," albeit with the same end in view and completed—fully stocking their respective territories, that is. But then Darwin's own eyes lit on a lion-ant that showed, through its predatory habits especially, that it was, although specifically distinct, cogeneric with the European lion-ant. Faced with this creature, even the Disbeliever would have to concede, Darwin reflected, that one hand had worked throughout the universe. However, reconciling the contrast between the two faunas with the singularity of God still left their distinctions unexplained. And Darwin ended by writing that "a Geologist perhaps would suggest that the periods of Creation have been distinct and remote the one from the other; that the Creator rested in his labor." It is reasonable to read this comment as showing that Darwin has now decided that geology, even Lyellian geology, cannot explain how the Australian mammal fauna contrasts with every other. He knew that Europe had had marsupial species as far back as the Secondary ages, well before all the placental species now extant had originated. So why had Australia had none of these placentals originate in it? Lyell or anyone else notwithstanding, conditions there were no answer; witness how well the European placental species did as soon as they arrived. What, then, would be gained by arguing that the stocking of the world with mammal species was done twice and at very remote periods? Suppose that the whole world was given marsupial species the first time; it has still to be explained why, after a long rest, everywhere except Australia was later given placentals the second time. In any case, a Lyellian's God does not ever rest, in that he is always and everywhere bringing new species into being.[94]

Even if Darwin was not reasoning along precisely those lines, it is evident nonetheless that he was confronting a biogeographical explanatory challenge that he understood to be distinct from those he had faced in South America. As long as he had not been discriminating in certain ways, certain difficulties did not have to arise. A low total number of indigenous species in a region could be credited to the youthfulness of the land, provided that, as so often in South America, there was indeed geological evidence for that youthfulness. But when what is so strikingly low is not the total number of species, but the number of species of certain supraspecific groups (mammals generally, for example, or marsupials), and if the land (as in Australasia) is obviously not youthful anyway, then another and a more discriminating explanation must be sought. Adaptation to conditions could have provided such a discriminating explanation, of course, if only the flourishing of alien colonists did not discredit it so directly. Where, then, to look, adaptation to conditions not being allowable? Well, Lyell had made much of the fact that different kinds of animals had different migratory powers; and that mammals having the least migratory mobility provided the best indications of zoological provinces. It is reasonable to suppose that his Australian findings and reflections first brought Darwin to seek, in the differential migratory powers of animals of different supraspecific groups, the explanation for their differential representation among the indigenous species in various regions. If so, then it is reasonable to suppose that it was those findings and reflections that first prompted him to consider the transmutation of species. For differential migratory powers can only provide explanations for such differential representations on the assumption of transmutation. Of course, the documents from Darwin's Australian weeks contain no direct trace of any such thoughts. There are circumstantial indications, however, that such thoughts were in play within months if not weeks of the Australian visit. They are to be found in the *Ornithological Notes.*

The weeks sailing between Australia and the Keeling Islands in March were probably used by Darwin for geological writing. We know, from his letters home, that he was working at this between Keeling and Mauritius in April, and again between St. Helena and Ascension in July. He would, perhaps, have mentioned the ornithology if he had been doing that too. So there is reason to think that that may not have been fitted in until August and September. Some such tentative dating has support from the discussions of Barlow and of Herbert. Barlow cites an entry in the zoological diary dated by Darwin at April 1836. It relates to the two ostrich (*Rhea*) species. Barlow reads it as revising what Darwin had entered in the *Ornithological Notes.* But, in fact, there is no direct contradiction, and so this April note could well have preceded the earliest *ON* entry on the *Rhea* species. Moreover, Darwin would probably have included the dated April note in the *ON* if it had then existed. There is also some slight support for dating the *ON* after Darwin's visit to the Cape (in June); it has two references to William Burchell's book on Southern Africa, a work he would be unlikely to have read far in advance of visiting that country. Herbert has shown that there is every reason to think the *ON* completed before the voyage ended. So the time between Keeling (April)—even more so Ascension (July)—and England seems the most likely moment for its later sections to have been written.[95]

There are two keys to understanding the telling Galapagos Island section of the *ON.* First (as with the dog that did not bark in the Sherlock Holmes story) we have to notice how Darwin does not explain what he finds conspicuously absent from the islands. And second we have to notice how one passage, albeit invariably overlooked, provides the decisive context for another, the most famous one.

Opening his five pages on the ornithology of the islands, Darwin notes their location and their entirely volcanic constitution. He then passes to climate and vegetation. Being arid, the islands are mostly, he says, thinly clothed with nearly leafless and stunted brushwood or trees. Between one and two thousand feet up, however, on the windward side, there is rain enough for the soil to provide a green and fairly "luxuriant vegetation."[96]

"In such favourable spots, and under so genial climate," he recalls, "I expected to have found swarms of various insects; to my surprise, these were scarce to a degree which I never remember to have observed in any such country." And how would he now explain this scarcity of insects? "Probably," he continues, "these green oases, bordered by arid land, and placed in the midst of the sea, are effectually excluded from receiving any migratory colonists." And he follows this conjecture by passing from the insects to the birds: "However this [the scarcity of insects] may arise, the scarcity of prey causes a like scarcity of insectivorous birds and the green woods are scarcely tenanted by a single animal."

So much for what is missing. "The greater number of birds haunt, and are adapted for, the dry and wretched looking thickets on the coast land." There, thanks to a short rainy season, plants flourish briefly each year to leave abundant seeds buried in the cindery soil. "Hence the Finches are in number of species and individuals far preponderant over any other family of birds."[97]

Now, taken together, the explanations for the scarcity of insects and for the abundance of finches only make sense on the tacit assumption that any indigenous terrestrial bird or insect species on these islands would have arisen there by transmutation of species following migration of ancestral species from the mainland. For Darwin is invoking migratory inaccessibility to explain why, by contrast with birds as a class, the insects are hardly represented by any indigenous species. If birds can be numerous there, because finch species have arisen there, why not likewise with insects? His migratory inaccessibility explanation must presuppose that indigenous insect species could not have arisen upon the islands independently of any earlier species already existing elsewhere in the world. The absentee families among the birds—the warblers say—are missing because no species of those families could flourish in these conditions, even if they could get there; that explains why there are no warbler species, indigenous or otherwise. But for the absence, or nearly so, of the whole class, insects, no such explanation is possible; because the conditions are perfectly suitable. Migratory inaccessibility for insects, without the assumption that the new species arise by transmutation of earlier ones, can only explain the absence of nonindigenous insect species. Without that assumption, and given that the conditions are perfectly suitable for insects and have indigenous bird species, the islands should be stocked with insects of indigenous species.

To see why this transmutationist assumption has to lie behind Darwin's reasoning here, consider what he is not considering. The islands are young land. Indeed earlier in the *ON* he had began writing of the Galapagos by noting that they consist "of a pile of recent Volcanic rocks."[98] Why, then, does he not mention this point here; why does he not explain the scantiness of this fauna as he did that around Valparaiso in 1834 and Banda Oriental in 1835? Why not say, following the precedent of Lyell on Sicily yet again, that there are few animal species on these islands because they are too young for many indigenous species to have originated there and too young and too inaccessible for many others to have settled there? Because, of course, as with New Zealand and Australia, he seeks to explain not only the low total number of species but also the lower number of species of some supraspecific groups as contrasted with the higher numbers of species of other groups. Youthfulness of land is too indiscriminate to provide an explanation. And the only discriminating explanations left for a Lyellian like Lyell, who rejects transmutation, would have to appeal to adaptation to conditions; but such appeals could not be reconciled with the fitness of the islands for what they lack. Darwin's not appealing to the youth of the islands or to the conditions there to explain what is missing suggests that he is now committed to transmutation. His appealing, instead, to differences in migratory accessibilities (for birds as contrasted with insects and mammals) confirms it.

His famous discussion of the mockingbirds puts it beyond a doubt. He begins by noting that these birds are "closely allied in appearance" to the Thenca of Chile or Callandra of La Plata; that their habits are indistinguishable from those mainland birds, although their cry may be slightly different. So far, then, nothing that Lyell, or Darwin in 1835, could not take in his stride. But now comes an emphatic generalization about the three distinguishable kinds among the four specimens collected: "In each Isld. each kind is *exclusively* found: habits of all are indistinguishable."[99]

Lyell had argued that a group of islands, some inhabited by species found nowhere else, would not mark an area of the world that had had more than its share of species creations. For, he said, an ancient island such as St. Helena (his own example) could have existed long enough to have many species originate on it over a great period of time, and then if several much more recent islands were to arise nearby they would receive in each case a different sampling of the species migrating as colonists from the older island. Now, like Wallace years later, Darwin is arguing that the Galapagos mockingbirds' distribution cannot be explained in this way. There is no island where more than one kind is found; so there is none that evidently served as a center from which they all once dispersed. Moreover, with their indistinguishable habits, they would presumably interbreed to make one kind were they ever all on one island together. How then does Darwin continue? When he recalls the Spaniards being able to tell which island a tortoise was from by the body shape, size, and the scales; and when he considers "these Islands in sight of each other, and [but *deleted*] possessed of but a scanty stock of animals, tenanted by these birds, but slightly differing in structure and filling the same place in Nature," he says, "I must suspect they are only varieties"; the only similar fact he knows of being "the constant asserted difference—between the wolf-like Fox of East and West Falkland Isds."[100]

Now, his zoological diary shows that he took the Falkland fox to be a distinct species from the several mainland ones; so, likewise, with the Galapagos mockingbirds; they are, he is presuming, specifically distinct from the mainland ones.[101] The three kinds, each on its own island, are only varieties in the sense that they are only local varieties. That is they have arisen as varieties in their present locations and are mere varieties, in that they are not different enough and presumably too inclined to interbreed to count as species. For how do all the considerations cited by Darwin, especially the closeness of the islands and their scanty stock of animals, support the conclusion that these birds are only varieties? Again, the argument is invoking the obvious implausibility of their having originated as so many distinct species elsewhere and then having all migrated so as to land up one each on each island. The islands being close, migration from one to another is fairly easy. But their scant stock of animals—recall the earlier passage—shows how very much more inaccessible each is from the mainland. So, put those two conclusions together and it is most likely that only one original mockingbird migration from the mainland has ever taken place, with several since from one island to the next. The slight differences on each island are then due to divergences that have taken place following migration from one island to another.

What, then, of Darwin's famous closing line: "If there is the slightest foundation for these remarks the zoology of Archipelagoes—will be well worth examining; for such facts [would *inserted*] undermine the stability of species"?[102] This reflection should be read as saying that such putative facts would, if confirmed in this case and matched in others, provide welcome evidence for the mutability of species; the "would" marking a twinge of conscience over the overly eager presumption that these putative facts are already beyond possible revision and are typical of Archipelagoes generally. But why should mere local varieties, rather than distinct species, on the islands of archipelagoes be evidence for transmutation? Because varietal divergence following island-to-island migration makes more credible specific divergence following an earlier mainland-to-island migration. Any single volcanic oceanic island, scantily stocked, and with peculiar indigenous species closely resembling others on the nearest mainland, will probably evidence transmutation in that it is likely to have a fauna which cannot plausibly be explained without that assumption. But island clusters with varietal differences between animals on nearby islands show, in addition, a smaller version of the migrations and divergences that have allowed such islands to have any indigenous species at all.

On this reading of the documents from 1835 and 1836, a decisive shift is seen, in what Darwin does with the Galapagos fauna on revisiting it in the *ON* that he did not do when visiting it on site. And that shift is tentatively traced to reflections upon a failure to reconcile the fauna of Australasia, when there, with Lyell's views as accepted by him in 1835. Further confirmation for this conjecture comes when we notice signs of Darwin's intense preoccupation, after Australia, with the presence or absence on islands of indigenous and nonindigenous species of birds, mammals, insects, and plants (Why not reptiles? Presumably, he was already thinking, as he would later, that islands such as the Galapagos had got their peculiar reptile species by transmutation of marine rather than mainland colonists). It comes even more strongly from the signs, as in diary entries relating to Keeling and St. Helena, of his preoccupation with the relevant contrasts in migratory powers and colonizing invasions. Plants, thanks to their seeds, are generally the most wide-ranging, and sure enough were represented at St. Helena by many European species that had very recently replaced many of the numerous species indigenous to that exceptionally ancient volcanic oceanic island. Plants would be, then, Darwin seems now prepared to argue (as he will remind himself to do in the Red Notebook) the first to be represented by new, indigenous species arising by transmutation of colonists on any new island. Among the birds, those of marine families such as the gulls range widest; waders perhaps next; and finches, warblers, and the like least. And, Darwin is apparently noticing, the representations of these groups fit well with these generalizations. So they fit well, too, with the supposition of new islands being stocked with indigenous animal species by means of migration and transmutation. Likewise for insects and mammals, these last being the least likely to cross ocean seas.[103]

On this reconstruction of Darwin's theorizing in these months, we also have a rationale for his intense interest, in the *ON,* in the character and distribution of the two *Rhea* species on the South American mainland. As both species were flightless, the southern one could obviously be assimilated to the attempt, begun in the reflections on geological notes in early 1835, to establish what exceptions there were to the generalization that the youthful land of Patagonia was stocked, particularly in its mammals, with species that had originated far to the north. But for Darwin's new generalizations, the southern *Rhea* was now an instance, not a counterinstance. As with the armadillos and cavies (and some further bird samples, noted in the *ON*) the southern *Rhea* species was congeneric with and so represented a northern species. So, well before the *Rhea* species—as inosculating (territorially overlapping) representative species—were (in 1837) providing evidence for species changing *per saltum,* as one mode of transmutation, they were, along with other representative species in the same continental region, providing evidence for migration and transmutation itself.[104]

However, even if the reconstruction conjectured here of the significance of any particular case such as the *Rheas* seems irredeemably uncertain, one basic conclusion can hardly be avoided: Darwin first went over to transmutation when, as a Lyellian historical biogeographer, he decided that he needed to add that explanatory resource to those he already had in play. Lyell had taught and Darwin, in 1835, had no reason to disagree that one could explain all the geographical facts with a suitable combination of a few considerations: the conditions and the geological history of the region and the adaptations and migratory powers of the species. By the end of the voyage, Darwin was disagreeing. That was not enough in many striking cases. No, one needed to consider the conditions and the geological history of the region and the adaptations, the migratory powers, and the transmutations of the species.

That Darwin first favored the mutability of species as an explanatory resource additional to these others is of major consequence for any understanding of his entire development as a zoonomical theorist in the 1835–38 period. To see why, we should consider two issues here.

First, adaptation: years later Darwin was to give up the notion that species are "perfectly" adapted. But that rejection is to be distinguished from his 1836 rejection of the Lyellian view that the adaptation of species to their stations is sufficient—when suitably supplemented by geological and migratory history—to explain their distribution. Kohn is therefore mistaken in thinking that Darwin's earliest biogeographical grounds for embracing transmutation required any change of mind about adaptation. As Ospovat has insisted, we must not confuse two things: Darwin's rejection of the assumption that adaptation is the sole determinant of when and where any species originates and his later rejection of the perfection of adaptation itself. After the first rejection and before the second (and so from 1836 until at least December 1838, when he was arguing that natural selection, as contrasted with artificial selection, could have produced the "ancient and perfectly adapted races" that species are) Darwin's position was that, yes, species are exquisitely adapted to their stations, as Lyell and others say, and any theory of the origin of species must explain how they come to be so; but no, all the characters of the species originating in any area cannot be credited to adaptation to the conditions there, some must be credited to heredity, to inheritance from older ancestral species there.[105]

Second, novelty: there have always been impressionistic clichés declaring that in Lyell's world nothing really new ever happens (an obvious exegetical error, of course; his whole chronology for the Tertiary formations explicitly presupposes that each new species has never existed before and, once extinct, never will arise again). And there have also long been clichés about "evolution" that play on that word's connotations of progress and innovation. So, not surprisingly there have lately been those (most notably Cannon) who would put these two lots of clichés together in concluding that when Darwin adopted mutable species he must have been repudiating his mentor's entire "worldview."[106] If such a conclusion is meant to be taken seriously, then it must be taken literally as a thesis about Darwin in 1835 and 1836. And so taken, needless to say, it could not be more mistaken. Most instructively so, indeed; Darwin's initial rationale for favoring transmutation was actually his conviction that the new species arising on new land could not be as new as Lyell held. What is happening at any one time is even more constrained by what has already happened than Lyell would have it. Even if conditions there have changed a great deal, from verdant forest to arid desert, the new species cannot have only new characters fitting them for the new conditions; they must also have some older characters inherited from their ancestors and not now adaptive, even if formerly so in the earlier conditions wherein their ancestors lived. And as for progress: in many of Darwin's earliest exemplary cases of transmutation, the *Rheas* and the quadrupeds of the American plains, the descendants were smaller than their putative ancestors, as Darwin emphasized himself (the words "change not progressife" and "degeneration," at RN 127 and 130, are telling here).[107]

Of course, Darwin's 1836 move to transmutation was not to be his last disagreement with Lyell over the laws governing the animate part of the terrestrial system, even if it was only his second and the first one after his 1835 move to Brocchian extinctions. It will be emphasised here, indeed, to a greater extent than anywhere previously, how very much more there was to come in the very next year, 1837. But it is always worth trying to avoid errors about the earliest parts of a narrative. So far, that thesis about the young protégé having to repudiate his mentor's "worldview" could only have led us entirely astray. Whether it must do so later as well, we can only decide after we have seen what did and did not change in Darwin's zoonomical thoughts, and their relations with Lyell's, in those extraordinarily productive months from the closing of the voyage (October 1836) and the opening of Notebook B with its "Zoonomical Sketch" (in July 1837).

## 9. Spring and Summer 1837: Going the Whole Hog Zoonomically

All the major innovations in Darwin's thinking in these months seem, in fact, to have come in the first half of 1837. And they are so remarkably comprehensive and consequential that they constitute a decisive *semiannus mirabilis* in his career as a zoonomical theorist. Much later Lyell and Huxley would use the vivid phrase "going the whole orang" of anyone who was prepared, as Lamarck had been, to subsume man in a general theory as to how new species arise in the modification of earlier ancestors. The phrase was presumably derived from "going the whole hog." We may then describe Darwin in these months as deciding to go the whole hog zoonomically. For sometime before July 1837 he took his fundamental structural and strategic decisions. Most especially, as we have already seen, he decided to develop two lines of argumentation. One took off from the contrast between sexual and asexual generation; and it ran all the way from the individual, adaptive variation in successive maturing ovules or eggs that sexuality makes possible, to long-run divergences as wide and wider as those between the Australian and other faunas. Here Darwin went the whole ovule. The second took off from the spontaneous generation of infusorian monads; and it ran all the way through higher and higher grades of organization to the mammals and man. Here Darwin went the whole monad.

There were, as Kohn especially has detailed, obvious and direct textual sources for the contrast that Darwin drew between sexual and asexual generation (in his grandfather's *Zoonomia*) and, equally, for his conjectured escalations of infusoria (in Lamarck's writings as presented by Lyell). So Darwin's decisions to respond positively to those sources, and to organize and pursue his theorizing in this way from now on, were of the greatest consequence not merely for the opening of Notebook B but for the rest of his life. With those commitments made he was going whole hog zoonomically, and that included going the whole orang anthropologically.[108]

The importance of understanding these structural and strategic innovations correctly will be obvious. Failure to do so can lead and has led to major misunderstandings of Darwin's intellectual biography. For, needless to say, if Darwin had embarked on some line of thought before the end of 1837, it will not do to trace his commitment to it in 1839 to some reading he did in 1838. We have recently been offered many strong claims, by Ruse, by Schweber, and by Manier, most notably, as to how Darwin's 1838 reading in or about Whewell, Comte, Adam Smith, and others prompted him to be thinking as he was about various subjects in 1839. Some of these claims are well documented, but—it is hard to be precise—perhaps some three- to five-sevenths of them cannot survive any exposure to the abundant evidence that Darwin was already committed to those views in 1837.[109]

It will also be obvious that we cannot appreciate the importance of these 1837 structural and strategic innovations if we see our task as one of plotting Darwin's position at successive moments by using two variables: the evidence he then has for transmutation and the mechanism he then favors for it. The evidence and mechanism for transmutation approach to the early Darwin would ensure that we sail right by all the interesting moments. And the fact that Darwin himself, in his various reminiscences (including that famous one about the Galapagos species being the origin of all his views), pioneered this radically misleading approach makes it all the more necessary that we should appreciate its inadequacies.

As befits a Piagetian with a fine sense of the structural developments in any theorist's intellectual life, Gruber has broken most thoroughly and fruitfully with this, Darwin's own retrospective approach to the Darwin of 1837. However, even apart from several important exegetical and analytic mistakes (some, but not all, since corrected by Kohn), Gruber has also—again, as befits a Piagetian—pioneered a new way to go astray in this territory. For he has apparently concluded—and certainly produced a sequence of diagrams entailing this conclusion—that the structural shifts in Darwin's early zoonomical theorizing were all strictly cumulative, so that each diagram in Gruber's sequence is derived from the previous one solely by addition (without any subtractions) of fresh elements and relations.[110] Not surprisingly, there is little textual evidence adduced for this schematic tidy-mindedness. In Piagetian hypothesizing about the stages in a young child's cognitive progress, one may be justified in making such Daltonian simplifying assumptions about the successive structures. There is no need to do so when we are dealing with an adult who has left us, in copious documents, plentiful indications as to which observations and arguments of his own and other people were decisive in leading him to think, to assume, and to infer, what he did.

We shall see shortly that the structural and strategic innovations—going the whole ovule and going the whole monad—decisive for Darwin's July 1837 Sketch came from a deliberate systematic, dissenting response made by Darwin to the whole sweep of argumentation that Lyell had presented in the *Principles* as his own response to Lamarck's doctrines. But we need, first, to understand how Darwin was then extending further those lines of thought already started by his 1836 disagreements with Lyell.

In Lyell's own zoonomical program questions about the laws of the animate system were answered as far as possible from what is observed today of extant species; and then one moved to the monuments of the changes recorded in those younger rocks whose fossils are of extant species, associated with an increasing proportion of the remains of extinct species as one goes backward in time. The geographical distributions of living species were thus indispensable, as the most accessible traces of the timing and placing of the origins of these species themselves. Moreover, given the uniformitarian premises presupposed by the whole program, they indicate—in that telling phrase of Lyell's—the laws regulating the introduction of new species; the laws for every past, present, and future change in life on earth, predictable and retrodictable from any monuments, geographical and paleontological.

Pursuing this program, from 1834 and 1835 on, Darwin in 1836 already had all the rationale he needed for making any new conclusion about extant species, such as their origins in transmutations, an entirely general conclusion about those laws. The generalization of this conclusion raised no further problems, then, any more than generalizing Brocchian senescence—to make that the law for species durations—had done. As long as there have been species living, there have been some perishing by senescence, and others arising by transmutation.

Once Darwin was back in England and cultivated avidly by Lyell in person, there was every reason to make explicit the association between his mentor and the quest for such laws. In July 1836 Darwin was privately reminding himself (at RN 52) to refer to the botanical geography of the Andes to the teachings of "Lyell. Ch XI vol II," the one that opened by relating the limits of botanical provinces to the age of mountain ranges and closed by consummating all Lyell's actualistic zoonomical inquiries with the conjecture of a constant, ubiquitous exchange of species. Within the year Darwin was explaining, to future readers of his *Journal,* that the striking botanical and zoological contrasts on either side of the Andes accorded perfectly with the great age of the range. For unless each species is supposed to be created in two places, the inhabitants should be, he said, no more similar than those in two countries separated by a broad strait of the sea. For, he explained, the many species that have become extinct will have been replaced by many creations of new ones, each confined to one side. Before he went on to recall finding many animals on the eastern slope specifically or generically identical with those in the desert plains of the Atlantic coast and so also Patagonia, however, he added a footnote; it pointed out (at Lyell's request, perhaps) that such faunal and floral contrasts, so interpreted, illustrate "the admirable laws, first laid down by Mr. Lyell, of the geographical distribution of animals, as influenced by geological changes." And he reflected that the "whole reasoning" is of course founded "on the immutability of species''; for, otherwise, the distinctness of the species on the two sides might be traced ''to different circumstances in the two regions during a length of time'' (*JR* 400).

Likewise, in explaining how his subsidence theory of coral islands can resolve many geographical distribution anomalies (by indicating where vanished lands formerly provided migratory facilities) he emphasized that here, too, he was illustrating those same ''admirable laws'' first established by Lyell. Again, his findings on the fossils and stratigraphy of the South American plains confirm, he told the Geological Society in May, that ''remarkable law, often insisted upon by Mr. Lyell, that the 'longevity of the species' among mammalia has been of shorter duration than among molluscs.'' And, needless to add, he had also long associated the law of the regional succession of peculiar types with Lyell's general law of the gradual exchange of new species for old.[111]

As Herbert has emphasized, Darwin's new confidence in these and other such pronouncements of law often depended on new expert taxonomic judgments, most notably by Owen and by Gould, far more authoritative than he could have made for himself in 1836.[112] On the geology of the places he had visited Darwin deferred to no one, but on the fossils and the bird and mammal specimens his identifications and descriptions had often had to be provisional; in many cases judgment was either postponed entirely or those assumptions were tentatively accepted that were implicit in vernacular names or explicit in local lore (often quite sound, as with the gauchos, as Darwin insisted).

Many evidential uses made, in July 1837, of voyage material depended on the new expert judgments as those had come in, especially in January, February, and March. Items 1*b, c,* and *d;* 2*a,* and 3*a* and *b* in our earlier inventory did so most directly. In this respect the famous RN (126–33) arguments seem to mark a point after Owen's but before Gould's findings were worked into Darwin's theorizing about species origins. As classified by Gould, the Galapagos birds, especially the finches and mockingbirds (the one reference is evidently to a mainland mockingbird species) are not explicitly mentioned, nor do they appear between the lines. But, as those very items in our inventory for July 1837 show, even when these judgments were taken into account they required no fundamental innovations, much less revisions, in the lines of reasoning deployed the year before in the *ON.* They were grist to a mill that was already rolling. Indeed, after welcoming Gould's conclusion that the Galapagos mockingbird specimens were of three distinct species, Darwin still featured them, in his July argument (B 7) as, like the tortoises, so many varieties.[113]

In general, and as our earlier inventory can now confirm, the immediate response—as in the case of the finches, no less than the *Rhea,* guanaco (llama) and mockingbird cases—was to see the new findings as reinforcing transmutation in two ways: by confirming that many supra-specific group representations in space and time are seemingly impossible to explain without accepting not only that new species can arise from old, but that many congeneric or cofamilial species can descend from a single ancestral species; and by confirming that these specific divergencies can be initiated *either* suddenly in changing circumstances without isolation *or* gradually on the isolation consequent on mainland-to-island or island-to-island migration. The Galapagos birds had apparently been associated with this gradualness since the *ON.* Lyell had argued that the fear of man is quickly acquired by species in the wild, and so exemplifies his thesis that prolonged changes in conditions do not bring continuing changes in structures and habits. It is very likely, then, that Darwin was already countering this argument in the *ON* (as he will in the *JR*), especially when he asks, regarding the Galapagos birds and their failure to become fearful of man, as yet: "Does the disposition or instinct of a bird gradually alter from any cause acting on *successive* generations?" So, apparently, and contrary to what Kohn and others seem to suggest, the mainland species formations of the famous RN passage were being added to gradual island species formations as presupposed in the *ON;* witness, too, how both modes are defended early in Notebook B.[114]

So, species mutability, species formation, gradual and sudden, and generic and familial divergencies from single ancestral stocks: such were the issues, in the spring and summer of 1837, as Darwin used the new findings to extend and consolidate his 1836 theorizing.

There was no way, then, that Darwin's decisions to go the whole hog zoonomically—to go the whole ovule, to go the whole monad—could have been taken or seen by him as merely two more such 1837 consolidations and extensions of his 1836 views made possible by the new 1837 findings about the voyage material.

The only way these new findings could have elicited those decisions—presuming they did so—was by convincing Darwin that the time had now come to take Lyell on, not only over the 1836 issues—fixed versus mutable species or common descent solely within species versus between species as well—but over all the issues, wide and narrow, that Lyell had associated with these, in the opening chapters of his second volume. (So, perhaps, in his oblique, cryptic, not to say self-serving and self-deluding, way, that was what Darwin was remembering in that famous recollection about March 1837, the Galapagos species and the origin of all his views; but by now we should all be beyond trying to figure out his thinking at any time from such throwaway one-liners. If Darwin could believe, as he wrote much later in his *Autobiography,* that his early notebook inquiries were not guided by any theory, then he could believe anything about himself in retrospect.)

We turn next, therefore, to Lyell's entire argumentation in those chapters. As we do so, we should note that for Darwin, in the spring of 1837,

Lamarck meant that author as presented there by Lyell. Although he may have met Lamarck's theories firsthand, years before at Edinburgh, it was apparently only in the spring of 1839 that he read him with care. Thus a list (quoted by Kohn) apparently of authors whose books were to be bought and including Lamarck's name, could well fit with that, as the date when Darwin read Lamarck for himself; for Murchison's name is on it, and his first book was due out then.[115]

We should also note that Lyell had deliberately dwelled on all the conclusions, including those on man, that Lamarck had been led to "while boldly following out his principles to their legitimate consequences." Lyell's own tacit intention in doing so was, in part, to bring out the danger in paleontological progressionism (à la Sedgwick et al.); by showing how, in the Lamarckian system, it was used to support an ape ancestry for man and so, by implication, materialism, conclusions manifestly as abhorrent to Lyell as to Sedgwick, but which Lamarck had not hesitated to embrace.[116] There was, then, an ironic corollary of Lyell's praise of Lamarck's boldness; any young protégé such as Darwin (or later Wallace) not deterred by those prospects, and with his own reasons for accepting transmutation as a theory of species origins, was, the mentor's intention notwithstanding, actually encouraged to emulate the Frenchman in pursuing all such consequences of that theory.

Lyell's first chapter divides into two stretches of argumentation (I and II), each divisible into two (Ia and Ib, IIa and IIb) subsections (these numerical and alphabetical designations are *not* supposed to map onto those used in section 7). Thus (I) considers various grounds, as adduced by Lamarck, for thinking species indefinitely modifiable in changing conditions. For (Ia) considers the case for variation between species, at any one time, being thought continuous; Lamarck's case being argued from the way character gaps between species get filled in as more and more material comes in to the naturalist from around the world. And (Ib) considers variation over time within a succession of individuals descending from a common stock, as indicated by the short run of human observation and historical records, and as inferred for the long run of geological change.

There then follows the transition (to IIa). It is made in a very instructive way that has no precedent in Lamarck's own writings. We presume, says Lyell, that all the varieties of any one species have descended from a single common, original stock. So, if the case for unlimited species mutability is granted, then all the species of a genus, even of a family, could have done so too. The questions thus arise, he says, from what stem in any such case, have the many diverse forms since "ramified"; and, further, whether there have been many or merely one ultimate ancestral stock from which all subsequent diversity has branched. Now, there being "no positive data"—direct observational testimony—to settle these questions, Lamarck has, Lyell reports, deemed two considerations as important guides to "conjecture." First, among animals there is one graduated scale of organizational complexity and likewise another for the plants. Second, the fossil record is taken to indicate that the main types of life have appeared on earth in a progression matching their organizational ranking. So, what Lamarck has done, Lyell explains (it was not so, in fact, as any reader of Lamarck knows) is to put his transmutationism together with this paleontological progressionism, to get a conclusion directly countering the "ancient dogma" of a degeneration in nature from her greater perfection when fresh from the Creator's hand. Inert matter has given rise to life, the insensate living to the sensate, and then, finally, the irrational to the rational.[117]

But how, then, Lyell now asks, does Lamarck cope with the two obvious difficulties for this scheme of progressive and continuous production: the persistence of the simplest sorts of organism; and the character gaps, the chasms, among the most perfect, that even Lamarck admits will never be filled by naturalists? He does so, we are told, by supposing, first, that the very simplest organisms, infusorian "monads," are daily produced in spontaneous generations; so the lowest grades are constantly refilled from below. And, second, in addition to the general tendency to advance (which would produce a completely continuous linear scale) all organisms have been influenced by external circumstances, as they have spread into diverse habitations themselves changing over time, as geology reveals. Instead of rising always along a single gapless line, life in its higher forms progressed along many branches, leaving character chasms as it went. Such, says Lyell, is the "machinery of the Lamarckian system." And he proceeds to show how it is supposed to have worked: in turning the simplest organisms into higher ones; and, in particular and in detail (IIb), how man is supposed to have come from the orang.[118]

The structure of Lyell's whole presentation of Lamarck's theorizing thus determines the structure of his own response to it; and so in turn Darwin's response to that. For, clearly, if the Ia and Ib conclusion of indefinite mutability is rejected, there is no need to have a separate rejection and replacement of the IIa and IIb arguments. And, sure enough, Lyell has two and a half chapters on the mutability issue as such, and only a few pages on transmutation as conjoined with progression, as in IIa and IIb.

Progression itself, he reiterates, in opening those few pages, cannot be established from the lack of mammal fossils in the known Secondary rocks, because they include no river and few lake formations and so no sure indication of the animals living on the continents of that period. The only well-evidenced case of a higher type of life arriving late is man himself, who was of a new family, even a new order, as well as being a new species. But while this lateness supports progression, it does not support progression by means of transmutation. Likewise for progression as evidenced not paleontologically but from comparative anatomy and comparative embryology. The graduated scale of skulls and facial angles among vertebrates is no indication of a tendency for species to transmute into higher ones. Nor does the way a developing mammal takes on the forms of fish and reptiles, below it in the organizational scale, secure transmutation, much less progressive transmutation of the lower into the higher species. So, even if progression is accepted on paleontological, anatomical, or embryological grounds, progression with transmutation is not thereby evidenced unless transmutation as such is already secured on other grounds.[119]

Now, Lyell's whole critique of these grounds had made his argument depend on the one decisive issue: what short-run variability in species, in the historic, human period, indicates as to long-run variability over the vast prehuman geological ages. And it had made that issue depend in turn on a providential teleological one. What does the variability that we observe in the short run do for species in the long-run? Why have they been endowed with it by the Creator? It was on this providential teleological question that Darwin would refuse to take Lyell's stand against transmutation; and, with borrowings from his grandfather, would take his own stand against Lyell.

To see how these issues arise in Lyell's critique, consider next how Lyell relates two prior issues to one another: first, the definition of the term *species,* together with the criteria used in assigning individuals to species, and the reality, the existence of species so defined; and second, their mutability, limited or otherwise.[120] On both issues, Lyell was explicitly asking whether Lamarck had succeeded in discrediting the traditional view (associated by Lyell with Linnaeus) in favor of his, Lamarck's, own.

The traditional view is best introduced, here, by examples that Lyell himself might have used. On definition, the trick is to see why lions and tigers traditionally counted as two species, while Swedes and Italians did not. A familiar trio of criteria is involved. First, a character (structures and habits) gap: that between lions and tigers cannot be filled in by any intermediates; not so with Swedes and Italians. Second, true breeding: lions never produce offspring more like tigers than themselves and conversely; not so Swedes, who sometimes have children who can pass for Italian and conversely. Third, noninterbreeding: lions and tigers are reluctant to mate, only rarely produce hybrid offspring, and if they do these are infertile when bred among themselves; not so Swedes and Italians.

So species are defined as those groupings of individuals that meet these criteria of character gap, true breeding, and hybrid sterility. For species, so defined, to exist is for those groupings to be really delineated by nature; rather than arbitrarily by man as constellations in the heavens are—to use Lyell's contrast. The reality of such groupings is important to Lyell, as a geologist, as he emphasises. He wants not only to assign to their species the mollusks living in the Mediterranean seas today, and so to determine how many species are present; he want also a species count among the fossils in the Sub-Apennine Tertiary strata on the nearby land, so that he can discern the age of these strata by establishing what proportion of the fossils are of extant and extinct species. Gradual mutability over time would undermine this chronological procedure by discrediting character gaps between living and fossil organisms as reliable signs of distinctions between extant and extinct species.[121]

Conversely, as Lyell reports, Lamarck would argue that the difficulties in assigning individuals to species on the traditional criteria make long-run mutability more credible. As knowledge increases, according to the argument, character gaps are often filled; one finds descendants deviating from parental characters in changed conditions, especially those brought about by man; and one finds parents of very different character having fertile hybrid offspring. These findings are thus continually casting doubt on all the species assignments already made. Doubt is cast on all distinctions as more and more cases come in of individuals initially separated by sharp gaps but now shown to be conspecific by all the usual criteria. So even in the human, historic period, the demarcation of species from species seems shifting and only arbitrarily made. Add, next, the geologists' apparent findings from fossils: organic structures differing so as to be less and less like the present and less and less complex as one goes into the remoter past, and physical changes accompanying organic ones down through the ages. Is not indefinite mutability and progression the only conclusion easily reconcilable with the difficulties of the naturalist and the results of the geologist? Or so anyone following Lamarck would say, Lyell declares.[122]

It is this inference about that conclusion that Lyell rejects, arguing that there is another inference possible from the difficulties of the naturalist, one that does not support that conclusion when the geologist's findings are confronted.

His argument does not start, therefore, by rejigging the definition of species, but with five zoonomical "laws" in the "economy of the animate creation." It is these laws, he says, which when assumed hypothetically can be shown to be reconcilable with all the relevant facts. The laws themselves are hardly remarkable: that an individual can be limitedly modified by changed conditions; that such acquired modifications can be transmitted to offspring; that there are fixed limits in the deviation of descendants from parental characters; that each species springs from a single initial stock and is never irreversibly merged with another through interbreeding; and that each lasts for a long time.[123]

What is telling is rather the further presumption that Lyell introduces, at once, as the decisive ground for his reconciliation; for he wants to show first that these laws entail that, given a graduated scale of life, some conspecific varieties should differ more in character than some other distinct species. For some species are more variable in character, because fitted for more variable habitations (especially in temperate regions); others are less so because fitted for less variable ones (in the tropics especially). And this greater or lesser variability in varying conditions is for each species an original constitutional endowment, provided at its creation, Lyell says, by an Author of Nature. He (God, not Lyell) has foreseen all the possible circumstances that all the descendants of the initial stock will have to cope with as their habitation varies over space and time throughout the prolonged duration of the species.[124]

The disagreement with Lamarck is, then, as Lyell insists, over the extent, limited or unlimited, of variability in the long run. And Lyell proceeds to argue his whole case as a defense of his five laws and this providential teleological interpretation of them. He takes domesticated species as his principal witnesses. For in these the variability is best known. They can then refute Lamarck's view that greater knowledge eliminates specific distinctions as illusory. Dealing first with their structural variation and then with their instincts, Lyell emphasizes throughout how remarkably variable they are. It is to be expected that they would be, seeing that these species were destined to accompany mundane man in his travels. But all this variation in a domesticated species arises, Lyell argues, not as the acquisition of new organs or instincts, but only of new modifications of those that originally allowed the species to thrive in its earlier, wild habitation. Thus does he go through those cases that Darwin will simply take over from him—the dogs, the cabbages, the primroses, even the wild birds now almost everywhere become fearful of man—documenting the full extent of the variation; but arguing, throughout, that it all arises in the responses of an original constitution to new conditions; that it is quickly acquired and inherited and cannot thereafter be extended by further changes in conditions, and so cannot lead to a new species by transmutation; any more than the infertility of interspecific hybrids (when bred among themselves rather than with the parent stocks) allows for new species to arise from old by hybridization. Thus, then, does Lyell uphold the reality and the limited mutability of species.[125]

Now, Darwin could take over so many of Lyell's examples of domesticated species variation precisely because he was not merely disagreeing with Lyell about cases; his disagreement was over the fundamental premises brought by Lyell to their interpretation.

Nor should this surprise us. There was one way Darwin, in the spring of 1837, could not disagree with Lyell over short-run variability, especially as shown by domesticated species. He could not do so on the ground of observational and experimental findings about the human, historic present known to him and not to Lyell, because almost without exception (the persistent tameness of the Galapagos birds is one) there were none.

No, what he did was to reject Lyell's premises, his providential teleology of variability; not because it was such, but, rather, to replace it with another, the no less providential teleology of variability by means of sexuality upheld by his grandfather. Thus are we brought to see why Darwin went the whole ovule in the spring of 1837.

The importance of his doing so for his entire career will now be plain. As Gruber and Kohn have shown, one does not have to be a vulgar Freudian, making a historiography out of prurience, in order to conclude that, as a zoonomical and "metaphysical" (mind, morality, and society) theorist, the Darwin of the early notebooks is intensely, overwhelmingly preoccupied with sexuality, with the contrasts, and their consequences, between sexual and asexual generation.[126]

Darwin's earliest interest in such contrasts, when at Edinburgh, had probably been reinforced by the reading that we know he did then in his grandfather's book (reading encouraged, perhaps, by his Edinburgh mentor Grant, who was a maverick supporter of Lamarck's views, one recalls). But, now, what was decisive was, of course, a much more immediate precedent: the view of all generation as division that he had worked out to go with Brocchian senescence as a theory of extinction. In that theorizing, he had sought to show that sexual and asexual modes of generation all involved a succession of individuals arising by division but limited in their total span of life. In that theorizing, therefore, he had had to argue that sexual was more like asexual generation than it appeared and was usually thought to be.

By contrast, now, he needed to understand how and why sexual could be so unlike asexual generation. Here Erasmus Darwin had not only indicated that this problem was the pressing one, but also suggested where its solution might lie. Erasmus had asked why sexuality is so prevalent, so that even hermaphrodite animals mate; and he had argued that its "final cause" was manifest in two considerations. First, in generation without sexual crossing, in grafting in his example, the offspring exactly resembles the one parent. So there is none of that capability for "change or improvement" made possible by sexual crossing, where the exact resemblance is missing. Second, sexual crossing between species has allowed vast numbers of new species to arise in the hybridization of a few original ones.[127]

This rationale for the second consideration Charles Darwin could leave aside, but not the reflection already offered in its support by Erasmus. That the final cause of sex is nothing less than the production of new species from old (by hybridization) is confirmed, the grandfather reflected, when we see that the formation of sexual instrumentation comes as nature's consummating masterpiece in the maturation of the individual's organization from the simple "filament" that every individual is at its embryonic origin; witness, most vividly, insects that go through earlier larval and pupal stages of maturation from an initial filament, stages without the sex organs, which are only formed in reaching adulthood.[128]

As Kohn has shown, Darwin's references to his grandfather's *Zoonomia* in the opening pages of Notebook B leave no doubt that these were the decisive arguments for the grandson.[129] Nor do we need to guess as to how he first saw them complementing the theory of generation as division that he had already worked out to go with Brocchian species extinctions. Darwin himself makes the connection explicit, although hardly lucidly so, in a passage at the end of chapter 12 of the *Journal of Researches,* one which—apart from a footnote probably added in proof later, in the summer—we may presume was written around March, April, or May of 1837.[130]

Reflecting once again on colonial polyps as "compound animals," he there argues that such colonies of individuals in intimate union together make sense when compared with the "known organisation of a tree" as comprising so many individuals—so many buds—belonging to a common body.

Citing examples he had cited in the Red Notbook (RN 131) and will cite again in the opening sentences of the July Sketch, he argues that in a polyp colony or a tree the individuals are not only incompletely divided from one another, they are "produced only with relation to the present time," and so, by implication, in relation to present circumstances and conditions. By contrast, then, in sexual generation "the relation is kept up through successive ages" by means of "intermediate steps or ovules." The corollary of this contrast is thus that asexual division extends life by dividing it, but extends it only for a "fixed period," while the division in sexual generation makes each life shorter than asexual generation could make it; for each new ovule is a new life. Now, the advantage from this starting anew is that characters that were never or are no longer adaptive can be left behind untransmitted, while the new ones needed to thrive in changed conditions can be acquired and transmitted. By sexual generation, with ovules, "many peculiarities" which would be "transmitted" by asexual division are "obliterated" and "the character of the species is limited"; it is kept from changing in unfitting ways; while, on the other hand, "certain peculiarities (doubtless adaptations) become hereditary and form races." In these two circumstances, then, Darwin sees "a step toward the final cause of the shortness of life."[131]

Looking ahead, then, to the July 1837 Sketch, we can now grasp Darwin's alternative to Lyell's providential teleology of species variability. Understand, he is saying, where the limited short-run variability manifest in the historic period comes from: namely, sexual generation. Then, one must conclude that the same means whereby the character of a species is kept from deviating beyond these limits, in the short run, also provide for unlimited adaptive divergence in the long run and so for the species to propagate itself by transmutation into another species, rather than ending without issue when Brocchian senescence creeps up on it. Short-run adaptive variability has not been conferred separately on each species, by the Creator, in giving that species its original constitution. Limited short-run and with it unlimited long-run adaptive variability has been conferred, by the Creator, on all species by endowing all life, even the simplest, with the power and the means of sexual generation.

To confirm that such was the rationale for Darwin's going the whole ovule, we need only to look to the opening of the July Sketch. But, before we do so, we should prepare ourselves to understand that Sketch in its entirety by considering here what was involved in going the whole monad.

One must not get too carried away by this subject; phrases such as "the monad theory of evolution" are especially to be shunned like the plague. Gruber started such misleading talk, because he got overexcited. Kohn then overreacted. He sorted out several major confusions in Gruber (especially the fundamental one whereby Gruber muddled up senescence for species taken one at a time and senescence for the entire issue of species arising from any monad). But then Kohn went on to conclude, incorrectly, that the whole monad business was no big deal for Darwin anyway.[132]

To see how and why it was important we must avoid being distracted by the word. Leibniz is not relevant here. The connotation of extreme simplicity (which was decisive for Leibniz) associated with the word's Greek root is, however. For Lamarck had deliberately used it for those infusorian organisms that he thought to be so simple in organization that they could not be simpler and still live. They have a quite undifferentiated organization, the bare minimum for maintenance of vital activity. Darwin's other synonyms—"living atom," "monucle," and so on—for "monad" often carry the same connotation. These synonyms were borrowed, almost certainly, from an article on infusoria and related topics by Ehrenberg as translated in Taylor's *Scientific Memoirs.* He probably had not read the article much before early July as it was not apparently available in Taylor's publication until then, although the relevant parts of it had already appeared in the *Edinburgh New Philosophical Journal.* It would seem, therefore, that the July 1837 arguments about monads had only been in play for a few weeks; and that they are unlikely to have come before the decision to develop the ovule theory of variability.[133]

Having followed the IIb part of Lyell on Lamarck, we can see how the extreme simplicity of the infusorian monads fits with Darwin's purposes in July 1837. For he then thinks (although he would not for very long) that he needs continual spontaneous generation of them, just as Lyell said Lamarck did, in order to explain the persistence of simple grades of organization in recent geological ages and on into the present. Moreover, beyond that use for them, Darwin, as we saw in an earlier section, wants to get a further explanatory dividend by investing them with a further property. Suppose, as he argues in an RN note and in the "Zoonomical Sketch," any monad and so the whole succession of species issuing from it has a vast but limited total lifetime; then those lines that have led to mammals must have changed in species most quickly; hence the shorter longevity of mammal as compared with mollusk species.

And that is really it; the whole monad business ultimately comes to nothing more or less than that. The complications come either from its connection with the whole ovule business or from its running into trouble and being given up by Darwin within a few days or weeks of writing the July Sketch. The connection with the ovules arises because, not surprisingly, Darwin wanted to accept the transmutationist corollary of the embryological law of parallelism as discussed by Lyell. A higher organism undergoes in its individual maturation a condensed rerun of all those changes of form that have been made over vast eons, starting with the monad origin of that line of life.

Consider, now, the other complications—arising in the rejection of continued monad production in constant spontaneous generations. Before Darwin silently decides against explaining persistent simplicity that way, he explicitly decides (B 29–35) that the vast but limited lifetime for whole monad issue must be rejected; for it has an unacceptable consequence: namely, the simultaneous extinction of all the congeneric or cofamilial species on one single branch of a tree of life. We have, then, to appreciate what he is left with after these two negative decisions. He is left with the production of some infusorian monads at least once, a very long time ago, indeed, way beyond the paleontological-stratigraphical horizon. He is left with the condensed recapitulation, in the individual maturation of higher animals, of the changes that have led to them from those first organisms. And (here one must disagree with Kohn) he is left with Brocchian senescence as his theory of species extinctions.

Why, then, should one make a fuss about Darwin's going the whole monad around July 1837, if he went on to change his mind so importantly and so quickly? First, because it was itself part of going the whole hog zoonomically in disagreeing with Lyell; second, because these two rejections early in Notebook B meant that a great deal of thought had to be given, and was, in the rest of that notebook, to explaining both progress and the persistence of simplicity, without either an invariable tendency to progress or a constant production of simple organisms by spontaneous generation; third, because it was in going the whole monad that Darwin first went the whole orang; and, fourth, because even with those two rejections made he could not and did not go back on that momentous decision.

## 10. The "Zoonomical Sketch" of July 1837 and Its Sequel: The Rest of Notebook B (July 1837 through February 1838)

We are now finally in a position to understand not only why Darwin's Notebook B begins as it does, but also why the entire argumentation of its opening twenty-seven–page sketch is structured as it is; and, moreover, why we should not think of this notebook as a "transmutation notebook" at all. We should think of it as Darwin himself did when he put the word *Zoonomia,* underlined, across the first page as his heading for the whole notebook. This word meant for him, then, not his grandfather's book, but what he took his grandfather to have had in mind in taking that as his title: the quest for laws of life, the same quest Lyell had undertaken in inquiring into the laws of the animate part of the terrestrial system. As Darwin would say in an inserted entry, but one made apparently at the end of 1837, when reflecting on all the programmatic prospects opened up by his theorizing so far: "The grand Question which every naturalist ought to have before him when dissecting a whale or classifying a mite, a fungus or an infusorian is 'What are the Laws of Life'" (B 229).

The aims, then, are zoonomical in a sense that should need no further elucidation. And what structure was articulated, what strategy deployed in pursuit of those aims? The answer lies in seeing how those twenty-seven pages not only followed decisive precedents in Lyell's presentation and critique of Lamarck, but also set, in turn, decisive precedents for Darwin himself in the rest of that notebook, and beyond that the rest of his career.

We can appreciate now that Darwin's July Sketch divides into two movements (I and II) for reasons that had their precedent in the division into two movements (I and II) of Lyell's presentation of Lamarck. Thus Darwin's movement I (B 1–15) concerns "change" independently of "progress"; for it concerns adaptive variability and diversification as made possible by sexual generation, and as making possible a single, common ancestry for the many distinct species of a genus, a family, even a class. The transition (B 15–17) then makes the shift from "change" to "progress." Finally, the second movement (II), following the precedent of Lyell's II*a* (and II*b* by implication) sets out to use continued infusorian monad production to reconcile progress with the persistence of simple organization, and to use extinctions and branchings to reconcile continuity of progressive change with character gaps among the highest organizational types.[134]

Darwin's principal preoccupation in the rest of Notebook B arises, therefore, from a striking asymmetry. For the argumentation of his movement I all holds up in his eyes for the balance of the year; so well indeed that it receives no reappraisal, only reinforcement with further complementary arguments. By contrast, as we have already noted, the movement II argumentation runs into deep trouble even while it is being set down; so new solutions have to be found and elaborated for the problems it concerns. Moreover, in this reconstruction work, in Notebook B after the Sketch, the main resources for handling these movement II problems come from the Sketch movement I. The outcome is, therefore, that the July separation of the businesses of those two movements is itself discarded and replaced. The initial separation between ovule succession (in sexual generation) as a theory of adaptive change and monad succession (in spontaneous generation) as a theory of organizational advance is lost. For, as Darwin now sees it, the theory of adaptive change can be made, with suitable supplementation, to cover organizational advance as well, without any continual monad production at all.

This result is of great importance for Darwin: and so for us in our efforts to understand the transformations undergone for the structure of his argumentation in this and subsequent years. For what is going to become decisive is not the redundant division between July movements I and II; what will be decisive is a new division between everything up to a certain point in movement I and everything else, from I or II, that comes afterwards. For by that point (B 6–10) in movement I Darwin has argued his way through to the adaptive formation of new species, not merely new varieties, from earlier ones; while after that point he goes on to confirm this account of species formation by considering how common ancestries and divergent descents can explain all sorts of facts, especially in historical biogeography. Now, when, much later, he is working out how to argue for natural selection as a theory of adaptive species formation, he follows—as his *Sketch* of 1842 shows so plainly—precisely the precedent set, here, by this sequence up to and beyond that point in the July Sketch movement I. For the *Sketch* of 1842 has two principal divisions: the first argues for the existence of natural selection and its ability to produce new species from old; the second argues that all sorts of facts, in biogeography, embryology, and so on, confirm its responsibility for extant and extinct species, for these facts are most easily explicable on that theory.[135]

To confirm this view of the July Sketch, it will be best to trace its argumentation from the start, with special attention to that point in movement I, to the transition from movement I to II, and to the trouble Darwin has in arguing his way consistently through to the end of movement II.

The opening contrast of sexual with other modes of generation depends on maturational stages (recapitulating monad progressions) in a product of sexual generation. Maturation provides for adaptation "to *changing* world" presumably, Darwin argues, because embryonic unlike adult organization is susceptible to modification. Maturation also provides for the elimination of accidents and mutilations, for they are not transmitted when the new ovule makes a fresh maturational start. Why, then, with "this tendency to vary by generation," is any species so "constant" in character throughout its geographical range? Answer: another distinguishing feature of sexual generation—two parents—makes it possible, thanks to the "beautiful law," the providential legislation, that offspring are intermediate in character between their two parents. Now, these very features of sexual generation that ensure constancy of character, in a species living in conditions only fluctuating locally, can also allow unlimited changes when conditions alter, overall and permanently. All that is needed is for the conservative action to be circumvented. Now a few individuals separated from the rest on a fresh island would undergo the same modifications by the changed conditions and so, with like breeding with like in a small population, would not be diluting differences in crossing, a possibility confirmed by cases such as the Galapagos mockingbirds; and, likewise, there would be no dilution from crossing, if, without island segregation, a variety could be suddenly produced by changing circumstances and be disinclined to continue interbreeding with the present species, a possibility confirmed by cases like the *Rhea* species (B 1–9).[136]

Thus is Darwin brought, by his arguments for his new providential teleology of short-run and long-run variability by means of sexual generation, to the issues raised for species formation by the traditional criteria for specific distinctions. It is here, accordingly, that he acknowledges that he must confront Lyell on Lamarck on character "gradation," and "Non-fertility of hybridity etc., etc." (B 9–10). That confrontation finds its place, therefore, after the variability argumentation and before the historical, biogeographical argumentation. For Darwin next pushes on to confirm "this idea of propagation of species," new species generated from old, with the Galapagos and Juan Fernandez faunas and eventually with the Australian mammals. So, that confrontation marks the place where the divide will later come between the two divisions of the Sketch of 1842.

But now, in this July 1837 Sketch, it is the presuppositions required in explaining major faunal differences, such as Australia shows, as due to very long separation of land and divergences from very remote common ancestors, that prompt the transition to the whole monad business of movement II.

Lamarck, as reported by Lyell, had two principles: *"the tendency to progressive advancement"* and *"the force of external circumstances."* Darwin is steadfastly resolved to have only one such and it is to be a "tendency to change" not to "progress." His way of explaining faunal differences supposes, he says, that in the long run of geological ages, and physical changes, "every animal has tendency to change." This tendency, he admits, is "difficult to prove," and all he can do here is to support its "possibility" by recalling the *Rhea* case and also the evidence for the efficacy of isolation on islands in quickening change in species. Pushing on, he nevertheless insists: "Each species changes," and then asks at once, "Does it progress [?]" Is there a tendency in every species not merely to change but to advance? Thus does he embark on the issues of Lyell on Lamarck IIa (B 17–19).

Now although Darwin's agenda derives from that treatment of those issues, his own response could not proceed as had Lamarck's (as reconstructed by Lyell). For Darwin had already made moves in his movement I, and in his transition to II, that had no precedent in Lyell on Lamarck. For recall how Lamarck (as reconstructed by Lyell) had related his two principles to one another. A tendency to progress had been invoked to turn two or more graduated organizational scales into so many gradual production lines. Then ramifying, adaptive divergences—consequent on dispersions over a terrestrial surface variegated in space, in its conditions, and fluctuating in time—were invoked to explain what, namely, gaps, hesitations, and reversals in advances, this progressive tendency could not explain.

For Darwin, however, it had to be the other way round. His use of his tendency to change, in his movement I, to explain historical biogeographical facts (from the Galapagos and Juan Fernandez to Asia and Australia) had already required just such branching divergencies upon migrations (especially with isolation) into variegated and fluctuating habitations. So his movement II challenge was the converse of Lamarck's (as reconstructed by Lyell). In turning to explain organizational superiorities in the present as the product of organizational advances in the past, Darwin had to ask what further assumptions he needed to invoke, beyond that movement I tendency to adaptive, ramifying change.

His answers are worked out over a succession of entries, tightly and densely reasoned, that fill up the balance of his opening Sketch and most of the next twenty or so pages (B 18–47). Even Darwin himself had trouble following his thoughts here. Three times (twice on B 26 and once on B 36) he resorted, as was rare with him, to drawing a diagram. By concentrating on these very different diagrams, and the quite distinctive reasonings occasioning each one, we can learn a great deal about his zoonomical theorizing at this time and subsequently.

He begins by answering the question whether each species not only changes but also progresses by going to the top and bottom of the organizational scale: "man gains ideas" and "the simplest cannot help becoming more complicated." Presumably he is arguing that man can only change by gaining new ideas without losing old ones, and so is, as the cliché of the day then had it, a "progressive being"; while the simplest organisms have the only organization possible at that minimal level, and so must advance if they change at all. In any case, Darwin decides that progress has to be understood as what has taken life away from simple beginnings: "and if we look to first origin," he says, "there must be progress." Moreover, what holds for any simple beginnings presumably holds for all: "if we suppose monads are constantly [this word inserted but not later apparently] formed, ? would they not be pretty similar over whole world under similar climates and as far as world has been uniform, at former epochs" (B 18–19).

Any geological age is, then, assured a supply of these simplest organisms. And of their issue in any age, Darwin says: "every successive animal is branching upwards different types of organisation improving as Owen says simplest coming in and most perfect (and others) occasionally dying out." For example, testaceous invertebrates in the Secondary age, such as those of the genus *Teretabula,* have propagated recent, that is extant *Teretabula* species, but the *Megatherium* has no living descendants. The armadillos and sloths are then, Darwin argues, related to the *Megatherium* not as direct descendants, but rather as codescendants of still older types, some of whose branches have since died out. So, the generalization here is that the large and the small both trace ultimately to lowly stems; but some subsequent branches from such stems have soared to great heights before dying out with no issue, while others have crept forward to have more modest surviving offshoots.

The extinctions are necessary corollaries of the earth being both fully stocked with life and changing in conditions in any age: "with this tendency to change [in species] (and to multiplication [of species] when isolated), requires deaths of species to keep numbers of forms [these last two words inserted] equable" (B 20–21).

With this argument Darwin comes to face a very general challenge: how to reconcile two lots of fillings with the full stocking of the earth. For he has constant new monad escalations to fill up the lower grades of organization as fast as older escalations empty them; and he has lines changing and splitting without ending to fill up the places left by those species that are becoming extinct as other lines end without splitting. What is more, of course, both these fillings have not only to be reconciled with each other but with the character gaps among the higher types.

As for the change and multiplication by splitting of species, two sentences (one here at B 21 and the other three pages on, at B 24) reaffirm that these are to be understood as already indicated in movement I. The "changes" are, he says, not resulting (Lamarck is obviously the target here) from the "will of the animal," but "law of adaptation as much as acid and alkali," the law, he is implying, that determines how embryonic organization is affected adaptively by physical conditions. On the other hand the multiplications of species are made possible because aversion to intervarietal crossing fixes, makes irreversible, initial varietal divergences (with or without geographical isolation): "A species as soon as once formed by separation [i.e., mockingbird paradigm] or change in country [ostrich paradigm], repugnance to intermarriage [increases *deleted*]—settles it." And, sure enough, Darwin is about to go into the evidence for such intervarietal "repugnance" provided by Andrew Smith's reports to him of Negroes and whites at the Cape (B 32–34).

It is telling, then, that the analogy of a tree is first drawn to represent the balance between splittings and endings, and also to represent the constant maintenance of the full quota of species: "Organized beings represent a tree, *irregularly branched* some branches far more branches,—Hence Genera.—As many terminal buds dying, as new ones generated. There is nothing stranger in death of species, than individuals." This tree is, then, doing one new job: explaining taxonomic generalizations genealogically; and one old job (dating back, for Darwin, to February 1835), that is, explaining Lyellian providential replacements for species that are continually dying like individuals. The irregularity of branching is thus the explanation for the finding that the genera of any family do not all have a standard number of species; some, the result of much branching, have many more than others (B 21–22).

So far so good. But, as is his wont, Darwin now wishes to get all the explanatory dividends that he can from this prolific tree of life: by going first to its monadic starting point and then to the biggest branches from its main trunks. For it is here (B 22) that he conjectures a limited lifetime for a monad, and so its whole issue, as an explanation for the shorter longevity of mammal species. The explanation only works on the assumption that monads have been continually formed, and that some older monad issues have already lived out their vast but finite span of millennia, and so ended in mammalian extinctions. The explanation depends on treating a monad and its issue as one enormously extended individual life. With this explanation Darwin has, therefore, added one more level of limited, individual life to the other three already in play: ovules as buds (and so as individual members of a colony); organisms themselves; and species propagating and dying as individuals. Darwin is accordingly deploying the tree of life metaphor both because of the analogies between branching in tree growth and branching in the adaptive diversification of a genus or family; and because of the analogies between the growth of a tree as a bud propagation succession and the growth of a monad issue as a species propagation succession. And he has trouble, to start with at least, in keeping these two sets of analogies coherent and consistent.

To see why the trouble arises, we have to notice that he has two explanatory strategies at work here. In the first, similarities of character among different descendant branches are ascribed to their deriving a common inheritance from a single ancestral stem; with their differences of character credited to subsequent divergence. In the second, the closeness (or distance) between two groups, not in the character, but in the complexity, of their organization is traced as to their having progressed a more or less similar distance along a line starting at the extreme point of monadic simplicity. The trouble comes, then (at B 25–26), when the conclusions from these two very different sorts of tracings conflict.

Having accounted for the short lives of mammalian species, Darwin conjectures that there might be a tendency in any group to diversify into six subgroups. His argument supposes that a class may give rise to two lots each of terrestrial, aquatic, and aerial species. And he apparently hopes to argue that if five of these six subgroups can be gathered into a circle, leaving one to go in the next circle, then he can explain why the Quinarian system of classifying (each group subdivided into five) works when it does (see also B 46).[137] However, if "each main stem of the tree is adapted for these three elements," land, water, and air, there will be "points of affinity" between "each branch" on that stem (B 23–24).

Now, Darwin means here by affinity what he and others will often call analogy. For he continues this line of reasoning by asking whether he need "not think that fish and penguins really pass into each other[?]" Aquatic birds do not have to have come directly from fish stock, he is deciding. Penguins resemble fish from their common but independent adaptations. And yet surely birds and fish as two whole groups trace to a common ancestral source. To accommodate these two conclusions, Darwin distinguishes living, visible joinings and passages from dead ones no longer to be seen: "The tree of life should perhaps be called the coral of life, base of branches dead; so that passages cannot be seen." But in this thought, it seems he has a further "contradiction" of the "constant succession of germs in progress," of the continuity of sexual ovules in progressive changes (see *JR* 212); this seemed contradicted, first, by the lack of a passage from fish to birds through penguins and, now, by the lack of a visible passage from monads to birds.

A diagram can, however, take care of these difficulties. It shows no fewer than five stages in the tree of life growth: a mere point; a line (all living); a forked line (all living); a forked line with a dead (dotted) stem; and finally a line forked into three with each fork only joined by dotted (dead) lower portions to the common dead (dotted) stem. A further diagram can then explain why "fish can be traced right down to simple organisation.—birds—not." For while both fish and bird stems branch out from a common simple source, the bottom of the bird stem only is now dead (dotted). So a further dividend from this coral of life closes the Sketch. More perfect organization, as in the mammals, has been reached through short species lifetimes and so more dying at the bottom of the branches; with the consequence that in the higher types a natural "arrangement" of species by affinities would put them on circles around empty, dead centers, while among "lower classes" one would have "a more linear arrangement." The laws of life are the same for all, but in their consequences there may be one rule for the higher, another for the lower classes. Or so it appears at this moment in Darwin's reasoning (B 23–27).

It was not long to be so, however. Within a few pages a radical rethinking had called for a new diagram (B 36) and a further run of some ten pages working out its immediate corollaries (B 37–47). The rethinking represents a taking over of comparative taxonomic problems by the genealogical (ancestor-descendant relationship) reasonings in the Sketch, as distinct from the gradational ones. For what Darwin now decides is that the species dying (twig deaths), requisite to balance species splittings, can explain all the gaps and so lacks of passages that we find. There is, then, no need to ascribe these to stem deaths, as in a coral of life rather than a tree of life. His reasons for this conclusion see him moving away, therefore, from any contrast between the bottom and the top reaches of the tree. For it becomes irrelevant whether one is a long way up from minimal complexity or not. What has become relevant is not how far one is, in any case, from the bottom of the organizational scale and so from an infusorian beginning; but how far forward one is from the common ancestral stocks whence have diverged the groups one is considering. Gaps and missing passages are thus explained by reference not to more or fewer dead branches as one comes up from lower to higher grades of organization; but by reference to more or less remoteness in common ancestries as one moves across a wider or narrower spread of diversity among species. In a word, this is the moment where the monads and their progressive issues are dispensed with, except insofar as they remain as the remotest of all ancestors, and so the ancestors with the most diverse descendants.

The new diagram (B 36) to illustrate this rethinking is prompted by a historical, biogeographical, and genealogical refutation of limited monad lifetimes: if one grants "similarity of animals in one country owing to springing from one branch, and the monucle has definite life, then all die at one period, which is not case"; therefore, Darwin declares with underlining, the monucle does not have such a limited life. No; the extinctions of congeneric or cofamilial species, like the splittings that they balance numerically, must be spread over time. Darwin accordingly explains pictorially how, if one ancestral species has now thirteen (echoes of the Galapagos finches) species as its descendants, then a dozen species contemporary with the ancestral species must be now without extant issue. Equally, along with the branching from that ancestral species that produced the thirteen living twigs there must be many twigs that ended in extinctions. Following in Lyell's footsteps (he had compared the commemoration of species births and deaths by the fossilizing and embedding process with successive census-takers recording human births and deaths), Darwin now works out the human demographic and genealogical analogies for these implications of his new diagram. The analogies confirm that those implications include an answer—in the increasing numbers of species extinctions between branches as they diverge—to his old difficulties with gaps between birds and mammals and so on. All that is needed is to extrapolate further and further back, to the "father," the common ancestral stock, and beyond that to the father of several such fathers, and so on. For "the *greater the groups* the *greater the gaps* . . . between them.—for instance, there would be great gap between birds and mammalia, still greater between vertebrate and articulate, still greater between animals and plants." (B 42). It is for Darwin a moment of triumph. And we can understand why, if we concentrate not on the simplicity of his conclusion or on its familiarity to us, but on the intricate series of arguments by which he had reached it starting from an entirely different position.

With this new diagram and its associated generalizations elaborated, Darwin finds himself needing to make no major structural or strategic innovations in the rest of Notebook B. To see why it should be so we may consider how his theorizing proceeds, from now on, in relation to the leading general issues: adaptation; species formation; extinctions; affinities and analogies; gaps and passages; diversification and progression. To be sure his treatments of these are all being constantly revised. But they are also being constantly integrated with one another. And the bases of that integration are those provided by the theorising undertaken in the first four dozen or so pages of the notebook.

The overall shape of Darwin's work in this notebook confirms this interpretation of it. The explicit consideration of the most abstract theoretical issues raised in those pages is sustained for another five dozen or so (through about B 109, that is). There then follow some seven or eight dozen where Darwin maintains his preoccupation with them much more sporadically, and is often responding instead to ideas and facts as he comes on them through reading he is doing. After that, however, there is a run of some three dozen pages (opening the third century of pages) where those issues are again consistently to the fore and where he is elaborating his understanding of them and their implications with increasing confidence and enthusiasm. And this confidence and enthusiasm hits a peak of programmatic explication on two famous pages (B 227–28) where the prospects for new inquiries into life, mind, and man are identified in the closing of the year.

The shape of these entries over these six months from July to December makes sense when we notice that in the autumn Darwin seems to have had less time than usual for this notebook, most obviously when he was in Shropshire as he was for a whole month starting in late September.[138]

We may say, then, that as autumn gave way to winter he was able to return in a more sustained way to the abstract theoretical business so resolutely pursued in the summer. And we may seek to clarify how the winter continued what had been inaugurated in the summer.

The summer had seen the new theorizing associated with the new diagram developed in four ways especially. First, the understanding of resemblances and differences had been refined, by pursuing the general thought that the "condition of every animal is partly due to direct adaptation and partly to hereditary taint; hence the resemblances and differences, for instance, of finches of Europe and America, etc. etc. etc." (B 47). For this generalization provided in turn for distinguishing resemblances due to common ancestral heredity (affinities) and those due to adaptations (analogies); and it provided, again, for distinguishing past ancestral connections, between groups that have branched out from common stocks, and present mediations of gaps by groups intermediate in character.

Second, the understanding of progression had been transformed by concentrating on continual ramification in the diversification of successive ancestral stocks into many descendants throughout the entire period of life on earth. For this ramification implied that the persistence of simple grades of organization could have two sources: in the survival of lines that had never progressed to higher grades, and in the branching off, from those lines that have, of new lines that have (thanks, say, to parasitical ways of life) declined in organizational complexity (B 108). From now on, then, progression is to be explained as the outcome of what has happened in some branches while others have done otherwise. Progressions are not reconstructed by reading a graduated scale as the record of a production line. The reconstruction of a progression now takes only the usual branching map for diversifications and reads organizational gains rather than losses in the appropriate lines.

That map is, of course, to be read as a record of species propagations. And Darwin develops even more explicitly the analogies implicit in his providential teleology of sexual generation. Sexual generation does not allow an individual life, given limited extension by grafting, to go on as such. But it does allow it to give rise to another individual rather than dying without issue. Likewise, then, sexual generation allows a species to go on in changing conditions, but only to go on as another new species with a new but limited lease of Brocchian life (B 60–63).

The great virtue, then, of the species propagations, splittings, and dyings mapped in the third diagram is that the very long run wherein an entire family or order or class diversifies can be read as an extrapolation of the shorter run wherein a species diversifies into so many congeneric descendants. The diagram can thus be read as a map for what Darwin had been working with in his July movement I: degrees of difference in indigenous species as ascribed to more or less complete and prolonged separations of the land masses, and so more or less complete and prolonged preventions of transport and migration. In July, the shift from change to progress had required adding the monad theorising to the ovule theorising as applied to historical biogeography. Now, before the end of the summer, having added species extinctions to the general account of adaptive divergence and so to historical biogeography, there is no longer any need for such additions.

If we turn next, therefore, to the winter pages we find no novel resources, only further refinements in the deployment of the summer ones. We may start with adaptation. With this Darwin had done little since July, but his principal premise remains, that adaptation comes with the influence of physical conditions on immature embryonic organization arising in sexual generation.[139] As for species formation, the leading challenge here was to understand how interbreeding may cease between varieties newly diverged from one another. Darwin sees most hope in two lines of thought: that in plants and animals reversion on crossing indicates incipient intersterility, and that in animals aversion to crossing may be a correlate, in their instincts, of structural differences. Both adaptation and species formation involve the relation between mind and body broadly distinguished. And Darwin is thus intrigued by his grandfather's suggestion that the susceptibility of embryonic organization to change may be shown by the effects of parental wishes, at the time of conception, on the sex of the offspring (B 227).

Species formation involves extinction twice over. Two species may be formed when the varieties intermediate between two other varieties are extinguished, perhaps through failure to adapt in changing conditions. And, of course, species formations are balanced numerically by species extinctions. Furthermore, in the winter Darwin is still explaining the presence or absence, today, of groups intermediate in character between others, as due to the prevalence or otherwise of extinctions in those branches of the tree in the past.

On progression, he now makes explicit (B 204) that his theory of species propagations entails some progression, in that the presumption is that the common stock ancestral to any diversified class of species, such as the mammals, will not have been the "most perfect." Rather we must suppose it "intermediate in character," intermediate that is (Darwin seems to be implying) between the character of the two main branches, placental and marsupial, of the class. So, this reasoning, he says, allows for increases and decreases in complexity in the subsequent divergences for the ancestral stocks. Moving to progression between rather than within the great classes, he insists that it is "another question" whether the whole animal scale is being perfected by "change of Mammals for Reptiles." This change, he says, can only be "adaptation to changing world." And it is clear by this that he means that it can only have arisen as the outcome of the transmutations of species adapting to a terrestrial surface changing à la Lyell, as in his movement I in July. Thus he goes on here to resist Lyell's argument that placental land mammals may well have existed at the time of the Secondary formation. But he does not argue, here or anywhere else in these notebooks, for mammals arriving since then as an overall adaptation of life to a planet changing à la Sedgwick or Buckland.

So, we see that Hooykaas was right in emphasizing that for Darwin organic progression was not, as it was for Sedgwick or Buckland, the corollary of physical progression.[140] Darwin's break with Lyell over the late arrival of higher groups such as mammals has, therefore, not allied him with Sedgwick or Buckland or anyone who would explain organic progression as a fitting of life to a planet that has been cooling and calming from a molten fluid beginning. Indeed, Darwin sees as his great challenge to explain how the organic progression, that has given the earth its mammals only relatively recently, can have come from the adaptation of earlier species to changes on a strictly Lyellian terrestrial world. For this world is precisely, for Darwin, as Lyell says: a world in balance physically and so climatically in the indefinitely long run; although it is a world that has cooled overall (thanks to reversible land and sea distribution changes and perhaps reversible orbit changes too) since the Carboniferous. Likewise with man's late arrival, Darwin implies (B 208–9). That event was not an event different in kind from other progressive advances. It should therefore come under the same account of progression.

We are now finally in a position to see how those famous pronouncements of programmatic prospects arise from these earlier Notebook B moves. The prospects are divided into two lots by Darwin (B 224–29). First are those which opened up by ramifying transmutation of species, independently of conjectures about the "causes" of these transmutations. The second are those opened up by the conjectures he is making as to the "causes." And this second lot of prospects includes not only general ones, but also more particular ones when the general are applied to the case of ourselves, of man.

In the first, the ultimate prospect is establishing generalizations about structural changes by reconstructing ancestor-descendant relationships, and inferring from these which structures have passed into what others and which have not. These generalizations will not only guide classifications, they will, Darwin insists, guide speculation about the future. Thus, in a case of special interest to us, we can already see, he says, that fishes will never give rise to men in the future.

In the second, we pass, says Darwin, from "transmutation and geographical grouping" (i.e., distributions read as traces of migrations and transmutations) to try to discover "causes of changes." This quest leads, of course, first to adaptation and to the role of instinct and structure correlations, especially in aversions to interbreeding. Adaptation, again, leads to sexual generation and to the maturational progressions of ovules, and so to ancestries and structural progress, as reconstructed in comparative anatomy. It leads also to the study of ancestry and progress in instincts, Darwin emphasizes. And that in turn leads to "heredity and mind heredity, whole metaphysics," in the Scottish philosophers' sense of metaphysics, that is, the laws of mind (indeed of all faculties in animals and humans) morals and society. Furthermore, by putting all these general conclusions together, we can know what "we," that is we humans, "have come from and to what we tend"; for instance, we can predict whether the races of mankind will likely diverge further or rather merge from interracial marriages.

Most generally, then, the quest is for nothing less than the "laws of change," indeed the "laws of life," as the causes for changes having gone as they did in the conditions there have been. For the explanation for the laws of change will be sought in the most general properties, actions, or tendences that mark off the living from the lifeless.

## 11. Retrospect and Conclusion

Our narrative business now completed, it will be as well to look back, with apologies for repetitions, at the main landmarks.

We began with Darwin halfway through his *Beagle* years, midway through 1834, leaving eastern for western South America for the last time. He was by then a zealous convert to Lyell's geological science. And what we had to grasp was that Lyell's geology included a comprehensive account of the laws regulating the animate part of the terrestrial system. For Lyell, and so for any protégé of Lyell's, geological science included zoonomical inquiries. And the zoonomical explanatory program in Lyell was a new one; it had had to be so, because it arose in Lyell's attempt to establish what no one else had thought to construct: a uniformitarian, actualistic reconciliation of a neo-Huttonian system of the earth with the faunal and floral succession established by paleontologists since Hutton's own day. In this attempt Lyell had made faunal and floral succession the outcome of a gradual (one-by-one) species exchange; with adaptation to conditions the principle determining the timing and placing of species origins (in separate or independent creations) and so the geological and geographical representation of supraspecific groups.

We saw, then, that we could best reconstruct Darwin's successive steps away from his and Lyell's 1834 position as he, Darwin, made them: as, that is, a series of disagreements with Lyell. As luck has it—no historian has any right to expect such calendarical coincidences—the next three years saw one decisive break each: 1835 was the year Darwin broke with Lyell over extinction, for he then adopted the view of Brocchi (that species like individuals die of old age) as respectfully rejected by Lyell;

1836 was the year he broke with Lyell over the separate, or independent, creation of species, for he then adopted the view that new species arise in the transmutation of earlier ones. This came when Darwin decided that adaptational considerations alone were not adequate to explain the timing and placing of the origins of species, and so not adequate to explain certain biogeographical findings. Then, in 1837 (indeed probably around March of the year, as perhaps also with the previous two breaks, it seems) came a decision to break with Lyell in an even more extensive way. For now Darwin decided to "go the whole hog zoonomically," as we saw it made sense to say. He decided to pursue structures and strategies of transmutationist zoonomical theorizing as comprehensive as Lyell had presented Lamarck's as being. This decision involved "going the whole ovule," that is, elaborating a theory of how sexual generation (with its successive ovules) makes long-run adaptive variability possible; and it involved "going the whole monad," that is, tracing even the most complex types of organization to lowly origins in the simplest possible organisms—infusorian monads. And going the whole ovule and going the whole monad involved "going the whole orang," that is, including man in a general theory as to how descendant species diverge from their immediate ancestors and so come, ultimately, from very remote infusorian stocks.

We saw, moreover, how that decision to go the whole hog zoonomically was acted on in Darwin's opening his July 1837 Notebook B as he did; for he opened it with a "Zoonomical Sketch" that comprises two movements: a whole ovule movement and then a whole monad one. And we saw how the structures and strategies of that Sketch set the decisive precedents for the rest of Notebook B, and so for the programmatic prospects for future inquiry that Darwin explicitly delineated at the end of that notebook and that year; prospects that were in turn to set the decisive precedents for all his subsequent zoonomical theorizing (including as an offshoot his "metaphysical" theorizing in notebooks M and N). Finally, we saw how Darwin's new explanatory program was still within that tradition initiated by Lyell in one fundamental respect; for to the Darwin of Notebook B as to Lyell, a principal challenge was to establish how far adaptation can explain the exchange of species, and so the faunal and floral succession, on a neo-Huttonian earth. For, where Lyell had argued that it can do so, Darwin was now committed to showing that it could only do so when supplemented with heredity, with ancestral inheritance in arboriform, ramifying species propagations. And, of course, to supplement Lyell's program in that way had required starting a radically new program of his own, as different from Lyell's as that had been from any before Lyell.

So we may ask here what conclusions there are, from this narrative, for our understanding of the origins and development of Darwin's earliest notebook zoonomical theorizing.

Some half a dozen negative conclusions seem plainly indicated for a start:

1. It would be well, if we could all agree to drive the word "evolution" from this whole area of scholarship. It was never there in the first place and and when injected retroactively only brings deep trouble. Face this fact, that one can see how truly bizarre it is that a writer, such as Mayr, should feel compelled to base his entire analysis of Darwin and his theorizing on a preliminary discourse on the many meanings of a word that played no part whatsoever in Darwin's intellectual life early or late. It is, likewise, even more bizarre that such a writer as Mayr, drawing on and seemingly aided and abetted by the best recent scholarship, can raise as his most urgent and fundamental biographical question: "when and how Darwin became an evolutionist[?]". It is inevitable that, having asked this thoroughly misleading question, he has to give it an appropriately misleading answer. Darwin's "conversion," according to Mayr, came out of his encounter with the ornithologist John Gould in March 1837, when Gould identified Darwin's Galapagos mockingbirds as three distinct species. That was "the watershed in Darwin's thinking," for the "destruction of the concept of constant species" came from that encounter. "Suddenly," Mayr declares, "everything appeared in a new light. What had seemed so puzzling about his observations on the *Beagle* now seemed accessible to explanation."[141] One can only remark how ironic it is that "evolution" and "evolutionism," as units of an antitypologist's historiographical analysis, should inevitably lead not merely to such totally implausible historical scenarios, but to such saltationist not to say special-creationist ones!

2. It would be well to remember that Darwin went on his voyage after being at the Edinburgh and the Cambridge of his day and not the Harvard of ours; that his principal reading was in Charles Lyell, not in Ernst Mayr or in Stephen Gould. The Harvard issues of the 1960s—populationism, allopatriotism, and so on—can make interesting entrées into Darwin's notebooks, as Kottler and Sulloway have shown. And the Harvard issues of the Goulden seventies—antiadaptationism and antigradualism—can also, as Ospovat and Sulloway have shown.[142] But it may still be worth asking what the issues were that Darwin thought to be at stake during and after the voyage, as he compared his collections and observations with Lyell's system and Lyell's account of Lamarck's system. Latter-day preoccupations with "the concept of species" and "the mode and tempo of evolution" map only very incompletely onto Darwin's concerns. A real watershed in his thinking, such as his going the whole ovule and the whole monad, must be overlooked if we make Mayr's disagreements with Goldschmidt, or Gould's with Simpson, our principal points of departure.

3. If we really do want to avoid misreading the early Darwin as the precursor of the later Darwin we shall have to be wary of the first of all

such misreadings, Darwin's own in his various reminiscences; we shall also have to seek alternatives to analyzing the later Darwin's theorizing into so many "ideas" or "isms," and to going on a quest for so many earlier moments of conversion or discovery wherein his thinking initially came to include these cognitive units.

4. We need to avoid a phony dichotomy between thinking of Darwin's 1837 zoonomical theorizing as being either a "worldview" or "merely" a "scientific theory" about the origins of species on the earth. All "worldview" talk is liable to mislead us, not suprisingly, given its Fichtean, not to say Hegelian, presuppositions. A worldview might reasonably be held to be one or both of two things. On the one hand, it might be a set of beliefs about the way the whole universe—stars, sun, earth, moon, plants, animals, and man—is arranged: for instance, in a closed, finite sphere with the earth stationary at the center; while, on the other, it might be a set of beliefs about the way the whole universe is made and is working: for instance, vortices of tiny passive particles jostling in a plenum in accord with a few laws of inertia and impact.

Now, clearly, in these two senses Darwin, in 1837, was not developing for himself or anyone else a new worldview. He has nothing to say about the ultimate constituents and operations of everything from here to infinity; and he leaves the stars and the solar system, even—indeed especially—the earth, arranged exactly as it was before; an oblate terraqueous spheroid, the earth is left orbiting steadily and stably in a slowly and slightly varying elliptical path around the sun, exactly in fact as it is found in Lyell and in John Herschel's and Mary Somerville's digests of Lagrange and Laplace.

So, was the Darwin of Notebook B not thinking big? Of course, it does not follow that he was not. He was thinking very big: not only going the whole ovule, the whole monad, and the whole orang, but the whole "metaphysics." Now, one might want to say that, even if a man's view of the solar system, for example, is not literally and directly touched by any conclusions from such an inquiry, surely his view of that view is, his second order interpretation of his first order beliefs. When Darwin first construed all his beliefs, including his astronomical beliefs, as recent achievements by reasoning apes adapting mentally to a distinctive bipedal way of life, then, one might want to say, that made a difference to all those beliefs themselves.[143] Yes, certainly. But that still leaves him with the solar system going about its balanced Lagrangean-Laplacean business as before. And (Schweber notwithstanding) that was the way it stayed for Darwin right on through notebooks A, B, C, D, and E (the odd brief remark or two on possible central cooling or nebular condensation not to the contrary).[144] When Darwin, Lyell, and Herschel talk of "astronomical causes" of climate changes, they mean long-run reversible orbit changes rather than—what none of the three was prepared to invoke as explanatorily indispensable—"cosmological" or "cosmogonical'' causes of climate change, that is causes tracing to a putative origin of the planet in, say, a hot molten fluid chaos or in an incandescent nebula. So, in sum, Darwin's 1837 theorizing added up to far more than a ''mere'' theory. But let us explicate how he could be thinking bigger than a mere theory without distorting our explications with misleading ''worldview,'' much less ''evolutionary worldview,'' talk.

5. The revisionist thesis of Cannon, although justly celebrated and influential, needs reevaluating at the least.[145] On Cannon's account Darwin's theorizing, especially in its understanding of adaptation and progress, resembled the ''Christian'' thinking of ''liberal Anglicans,'' such as Whewell and Sedgwick, more than it did the thinking of Lyell; even if those ''Christians'' were not among Darwin's immediate and direct sources.

Now, frankly, much writing in this vein seems in danger of fatal equivocation.[146] Is there something distinctively Christian about a brick, if Christians make and use it in building their physical edifices, or about a natural theology if Christians devise and deploy it in forming their intellectual structures? One may doubt it, seeing how intensely and explicitly natural theology, as found in Paley and in Butler (two orthodox Anglicans if ever there were any in their day) could be cherished and defended by someone like the young Lyell; for Lyell (eventally a Channing-Martineau Unitarian) was never a Christian, in the sense that he never accepted the scheme of the Creation, Fall, Salvation, and Redemption as set out in Paley's other famous book, on revealed theology, or in the second part of Butler's *Analogy,* the part on revealed religion. Darwin's early thinking about adaptation, providence, and progress owed as much to three apparent deists—his grandfather, Lyell, and Lamarck—as it did to any Christian. Indeed, he was still in 1837, if his *Autobiography* is accurate, more of a Christian than any of them. Insofar, then, as he was able to assimilate their teleology and so on, was that in spite of rather than because of his Christianity? Whatever we decide, let us not forget that one did not have to be a Christian to like providential interpretations of adaptation, progress, population, sex, and so on. Thanks to Cannon we are now wary of T. H. Huxley's entirely mistaken view of Darwin's relations with Lyell, with Whewell, and with their theologies, natural and revealed. But, it is surely plain that the new revisionism of Cannon and others now needs revising too.

In closing, it is clear that we can now sound two much more positive notes.

1. Historians of science have become acquainted with several cases of a phenomenon especially familiar to historians of philosophy: a Plato-Aristotle kind of relationship between one thinker and another. The relationship is distinctive because the one, although massively indebted to the other, yet departs knowingly and radically from his teaching. It has always been evident that the Ptolemy-Copernicus relationship is such a

case, even though Copernicus could not be Ptolemy's protégé in a literal, personal sense. And it has recently become evident that the Descartes-Newton relationship provides a no less instructive example. For what is important in these cases is not merely that the student was indebted to the teacher for a greater or lesser proportion of the conclusions reached in his theorizing, but that he derived from him his very starting points and his very understanding, initially at least, of what were the aims and structures that constituted their common endeavor. For Copernicus that indebtedness is obvious once one reads his book alongside Ptolemy's. For Newton all one needs to do is to ask, not how Newton may have concluded in favor of one or other of his laws of motion, but from whence came his prior conviction that inquiry into the principles of natural philosophy should be undertaken as an attempt to establish such laws. Ask this, and then one can understand how he could become the first Newtonian principally by being the first person to disagree with Descartes in all sorts of highly original ways, while pursuing, not rejecting, that systematic quest for a few universal laws of motion that Descartes had inaugurated.

To propose, then, that the Lyell-Darwin relationship also conforms to this pattern is to propose that Lyell's science not be seen as merely the most powerful among the influences at work in Darwin's zoonomical theorising in the later 1830s; it is to propose that it is an influence different in kind rather than degree from the others. For, having provided the initial framework, the agenda, and the priorities, it had determined how any other influences could be influential, how any other sources could be resources. Equally, of course, it is not to propose that Darwin's originality is one whit diminished. As the cases of Aristotle, Copernicus, and Newton all show in their different ways, for such a theorist to stand in such a relationship to another earlier one can be for him to be as original as ever a human mind has been.

2. In order to understand such a relationship we need units of analysis larger than the "theory," the "idea," or the "ism," for those are appropriate only when we are dissecting a body of conclusions or, to change metaphors, resolving a molecule of achievement into its atomic components. Now, several authors, writing recently on scientific change over the long run of history, have proposed that we must often work with units not only larger in size than these familiar units but different in kind. One thinks here especially of Kuhn, of Lakatos, and of Laudan.[147]

Kuhn's proposal, "paradigms," despite its suggestiveness, is best resisted by historians working on a topic such as the origins of Darwinian science. Apart from anything else—such as the invitation to project onto all ages and areas the sources of disciplinary identity and hegemony in twentieth-century physics—the paradigm notion is embedded by Kuhn, sometimes deliberately, sometimes not, in a muddle of "vision" metaphors and analogies.[148] The result is that much of his worldview talk is only acceptable to someone who has no difficulty with the assumptions made by those idealists, such as Fichte, who earlier invited us to construe judgment and reason in this manner. We have already seen that such talk can only lead us astray on our topic. There is, moreover, a fatal difficulty with the "worldview" as a unit in analyzing the origins of any enterprise such as Darwin's zoonomical theorizing. A worldview is an achievement, not an ambition, something one has, not something one can be intending and trying to have. A person's worldview is identified and characterised by what he accepts as solutions, and not by the problems he is attempting to solve.

It is, therefore, no coincidence that a historian hoping to understand the origins of an enterprise such as Darwin's early theorizing, and hoping to understand a relationship such as the Lyell-Darwin one, should find it appropriate to talk, with Lakatos and Laudan, of traditions, programs, and problems. And in using such terms here the intention has been to echo their proposals, albeit without taking a stand on the major issues that divide them. A Lyell-Darwin relationship cannot be understood through any analysis of "ideas" or "isms" or of "worldviews," precisely because what a Darwin gets from a Lyell is not principally conclusions, broad or narrow, or achievements, general or specific, but problems, presuppositions, priorities, and programmatic aims, intentions, and ambitions. That is also why such a relationship cannot be understood by trying to locate the two men by relating both to something that is identified and characterized by reference to a conclusion or a solution: "evolutionism," or "evolutionary biology," or "an evolutionary view of nature," or whatever. What is wrong with "evolution," for these purposes, is not merely the Whiggish presumption in identifying and characterizing the enterprise, the vision, or the doctrine by reference to our solution to the problems that we think are raised by organic diversity; but the way too many interesting questions are begged by using any solution for that identification and characterization.

The suggestion made here, then, is that in this case a historiographical preference for the one set of terms over the other is not merely a matter of how the historian chooses to interpret any findings once made; if this essay has succeeded in any of its sections, it has shown that these decisions really make a difference to what one does with datings, documents, exegesis, and narrative.[149]

NOTES

1. This essay owes a great deal to discussions with Peter Gautrey, Michael Ghiselin, Sandra Herbert, David Kohn, Larry Laudan, Rachel Laudan, Ernst Mayr, Robert Olby, Lou Rosenblatt, Sam Schweber, and Sydney Smith. It is a pleasure to have this chance to thank them.

2. C. Limoges, *La selection naturelle* (Paris: Presses Universitaires de France, 1970); and W. Cannon, "The Bases of Darwin's Achievement: A Reevaluation," *Victorian Studies* 5 (1961): 109–34. For the more recent literature, see J. C. Greene, "Reflections on the Progress of Darwin Studies," *Journal of the History of Biology* 8 (1975): 243–73; and the references in D. Kohn, "Theories to Work By: Rejected Theories, Reproduction, and Darwin's Path to Natural Selection," *Studies in History of Biology* 4 (Baltimore: Johns Hopkins University Press, 1980), pp. 67–170; in R. A. Richardson "Biogeography and the Genesis of Darwin's Ideas on Transmutation," *Journal of the History of Biology* 14 (1981): 1–43; and in M. Ruse, *The Darwinian Revolution* (Chicago: University of Chicago Press, 1979).

3. S. Herbert, "The Logic of Darwin's Discovery," Ph.D. dissertation, Brandeis University, 1968. See also Herbert's two-part article, "The Place of Man in the Development of Darwin's Theory of Transmutation," *Journal of the History of Biology* 7 (1974): 217–58 and 10 (1977): 155–227. On the relations between Darwin's geology and his biogeography see, further, M. T. Ghiselin, *The Triumph of the Darwinian Method* (Berkeley and Los Angeles: University of California Press, 1969).

4. See H. E. Gruber, "A Psychological Study of Scientific Creativity," in H. E. Gruber and P. H. Barrett, eds., *Darwin on Man* (New York: E. P. Dutton, 1974).

5. Kohn, "Theories."

6. 1982 is to see the state of Darwinian scholarship surveyed in a work of several volumes, and by many hands, edited by David Kohn.

7. H. Davy, *Consolations in Travel,* 3rd ed. (London: Murray, 1831), pp. 124–41; M. J. S. Rudwick, *The Meaning of Fossils* (London and New York: Macdonald & American Elsevier, 1972), pp. 101–217; P. J. Bowler, *Fossils and Progress: Paleontology and the Idea of Progressive Evolution in the Nineteenth Century* (New York: Science History Publications, 1976); M. J. S. Rudwick, "The Strategy of Lyell's *Principles of Geology,*" *Isis* 61 (1969): 5–33; M. Bartholomew, "Lyell and Evolution: An Account of Lyell's Response to the Prospect of an Evolutionary Ancestry for Man," *British Journal for the History of Science* 6 (1973): 261–303; M. Bartholomew, "The Non-progress of Non-progression," *British Journal for the History of Science* 9 (1976): 166–74; M. Bartholomew, "Lyell's Conception of the History of Life," Ph.D. dissertation, University of Lancaster, 1974; P. Lawrence, "Charles Lyell versus the Theory of Central Heat," *Journal of the History of Biology* 11 (1978): 101–28; and D. Ospovat, "Lyell's Theory of Climate," *Journal of the History of Biology* 10 (1977): 317–40. The analysis given here of Lyell's geology and biology will be developed much more fully in a monograph, now in preparation, by R. and L. Laudan and myself. The methodological aspects of Lyell's theorizing about climate and organic succession are discussed in detail in two articles, shortly forthcoming: R. Laudan, "Lyell's Place in the History of Methodology" and M. J. S. Hodge, "Darwin and Natural Selection: Roles for Methodological Ideals in a Theoretical Innovation."

8. Rudwick, *Meaning of Fossils,* chap. 3, "Life's Revolutions."

9. C. Lyell, *Principles of Geology,* vol. 1, chaps. 6–8 and 26. Unless noted otherwise, all references to the *Principles* will be to the first edition in three volumes (London: John Murray, 1830–33) as reprinted in facsimile, with an introduction by M. J. S. Rudwick, in the series Historiae Naturalis Classica, ed. J. Cramer and H. K. Swann. (Codicote, Herts: Wheldon & Wesley, 1970; New York: S-H Service Agency, 1970).

10. C. Lyell, *Principles,* 1: 92–143 and 2: 308–9; Ospovat, "Lyell's Theory of Climate"; and R. Laudan, "Lyell's Place."

11. C. Lyell, *Principles,* vol. 1, chap. 9 and vol. 2, chap. 4. See further, the studies of Bartholomew, Ospovat, and Lawrence cited in n. 7.

12. Adam Sedgwick, *Proceedings of the Geological Society of London,* 1 (1834): 187–236 and 270–316. See also Lyell to Sedgwick, January 20, 1838, in K. Lyell, *Life, Letters, and Journals of Sir Charles Lyell,* 2 vols. (London: John Murray, 1881), 2: 35–37.

13. C. Lyell, *Principles,* 2: 1–184.

14. C. Lyell, *Principles,* 2: 1–2.

15. C. Lyell, *Principles,* 2: 66–69.

16. On Lyell's two strategies see C. Lyell, *Principles,* vol. 3, chap. 1; and Hodge, "Darwin and Natural Selection." Rudwick is thus mistaken in arguing, in "Strategy of Lyell's *Principles,*" that Lyell abandons his actualism in tackling these problems about species replacements.

17. C. Lyell, *Principles,* I: 63–64, 88–89, and 2: 194–98; and L. E. Page, "John Playfair and Huttonian Catastrophism," *Actes du XIe Congrès International d'Histoire des Sciences* 4 (1967): 221–25.
18. C. Lyell, *Principles,* 2: 179.
19. C. Lyell, *Principles,* 2: 179 and 183–84.
20. C. Lyell, *Principles,* 2: 183–84.
21. The fullest account of the documents relating to the voyage years is in Herbert, "Darwin's Theory of Transmutation: Part I," and "Darwin's Theory of Transmutation: Part II." For the two scientific diaries, see H. E. and V. Gruber, "The Eye of Reason: Darwin's Development during the *Beagle* Voyage," *Isis* 53 (1962): 186–200. These diaries are mostly in vols. 30–38 and 42 of the Darwin Papers at the Cambridge University Library. The personal diary is printed in full, in N. Barlow, ed., *Charles Darwin's Diary of the Voyage of H.M.S. Beagle: Edited from the MS* (Cambridge: Cambridge University Press, 1933, Reprint ed., New York: Kraus Reprint, 1969). Extensive extracts from the field notebooks are given in N. Barlow, *Charles Darwin and the Voyage of the Beagle: Edited with an Introduction* (London: Pilot Press, 1945). The letters to Henslow are given most fully in N. Barlow, ed., *Darwin and Henslow: The Growth of an Idea. Letters 1831–1860* (London: Bentham-Moxon Trust, John Murray, 1967).
22. Darwin (not elected a Fellow until 1836) may have been present as a guest for Sedgwick's address. In any case, he would soon have been able to read it, when it was printed—either in *Proceedings of the Geological Society of London* 1 (1826–33): 281–316, or in *Philosophical Magazine* 9 (1831): 271–317. His autobiography recalls: "The present total oblivion of Elie de Beaumont's wild hypotheses, such as his *Craters of Elevation* and *Lines of Elevation* (which latter hypothesis I heard Sedgwick at Geolog. Soc. lauding to the skies), may be largly attributed to Lyell." N. Barlow, ed., *The Autobiography of Charles Darwin, 1809–1882: With Original Omissions Restored* (London: Collins, 1958), p. 102. On the significance of Sedgwick's alliance with De Beaumont, see M. J. S. Rudwick, "Uniformity and Progression: Reflections on the Structure of Geological Theory in the Age of Lyell," in D. H. D. Roller, ed., *Perspectives in the History of Science and Technology* (Norman, Okla.: University of Oklahoma Press, 1971), 209–27. For C. Lyell on De Beaumont, see *Principles,* 3: 148–49, 270–74, and 337–51. Inscriptions in Darwin's copies (now at the University Library, Cambridge) of Lyell's vol. 1 and vol. 2 show that he received the first as a present from Fitzroy, and so presumably before sailing in 1831, and the second in Montevideo, by mail, in November 1832. His copy of the vol. 3 carries no clues to its arrival in Darwin's hands, but letters to and from Henslow show that (despite Barlow's later dating) it reached him in the Falkland Islands in March 1834; although it was not read until after his return, on May 8 that year, from an expedition up the River Santa Cruz. See Barlow, *Darwin and Henslow,* pp. 77, 83, and 93.
23. Darwin's field geologizing in and around the La Plata pampas and the Patagonian coastal areas came mostly in four concentrated bursts of work: (1) at Bahia Blanca in September and October 1832, (2) extensive travels from August to December 1833, which began with a reexamination of the Bahia Blanca area, (3) study of the coasts during visits at Port Desire in December 1833 and Port St. Julian in January 1834, and (4) the Santa Cruz River expedition in April and May 1834. As Gruber and Gruber have shown ("Darwin's Development during the Beagle Voyage") Darwin's reexamination in 1833 of Punta Alta and other parts of the Bahia Blanca region prompted a much more positively Lyellian reconstruction of that region's history. Many decisive geological diary entries are on MS pages 62–74 of vol. 32 of the Darwin Papers, where Darwin, in entries added in 1833, rejects and replaces his conjectures of the year before. As a field notebook entry (for August 22, 1833) records, he first saw his new view as entailing that the mammal bones were remains of species that became extinct *before* there had originated the mollusk species now living in the ocean and found as shells on the plains. He was, of course, as Barlow notes, to change his mind on this point. See Barlow, *Charles Darwin and the Voyage of the Beagle,* p. 194. In May 1834 Darwin added many new entries to those made in January 1834 relating to the coast at Port Desire. An entry headed "1834 May Port Desire (Appendix)" starts: "Having read Mr. Lyell's 3 vol [i.e., vol. 3] respecting the metamorphic rocks; and having at S. Cruz seen proofs of several elevations; many facts are now much more easily explicable than they formerly were." Darwin Papers, 33: 243r. However, he nowhere implies that he is now having to reject an earlier understanding of the coastal geology. For

parallel remarks on Lyell in the letter to Henslow of July 24, 1834, see Barlow, *Darwin and Henslow,* p. 93.

24. In the entry in the geological diary—made during or very soon after actually being at Port St. Julian—Darwin does indeed conclude that the mammal quadrupeds lived on the coastal plain, but he does not go on to consider what this suggests about the causes of their extinctions: "In the place in cliff where the Argillaceous beds are cut through there was a group of large bones [here he writes, in the margin, the specimen list numbers 1722 . . . 1728 and 1736 . . . 1745]; the whole skeleton probably had once been there. The vertebrae were in a line joining on to the pelvis; the small bones of the extremity were in a group near articulation of main line [?] (1723).—This is interesting as showing the existence of large quadrupeds in Lat. 49°: 14′, in the Southern Hemisphere; in its geological age being so clearly marked.—The earthy [layer] formed a tolerably level plain on which were lying many Mytilus [mollusk shells] (2 species) with their blue colour, Patellae etc. and proving that it [the earthy layer] was at the bottom of the sea with great gravel plain.—Was this near the embouchement [?] of some river in the old continent (or first elevated land) when the large quadrupeds flourished? It is the first time I have seen any matter superior to the gravel." Darwin Papers, 33: 246v–47r. Cf. a similar passage in the letter of March 1834 to Henslow, where Darwin tentatively identifies the remains as that of a Mastodon. Barlow, *Darwin and Henslow,* p. 84.

25. See C. Lyell, *Principles,* 2: 130–57.

26. C. Lyell, *Principles,* 2: 158–68.

27. C. Lyell, *Principles,* 2: 169–73.

28. C. Lyell, *Principles,* 2: 169.

29. The note, included in Darwin Papers, vol. 42, was quoted in part by Herbert in "Logic of Darwin's Discovery," p. 52. Kohn has discussed it very extensively in his thesis, "Charles Darwin's Path to Natural Selection," PhD. dissertation, University of Massachusetts, 1975, and in part I of his essay "Theories." Kohn's and my own interpretation, reached independently, are encouragingly consilient, especially in relation to Lyell on Brocchi as a precedent for Darwin; although Kohn takes Darwin's talk here of the "existence of species" to mean immutable fixity of character rather than intrinsic limitation on duration. The comparison with Lyell's text shows that the latter must be correct; although there is, as we shall see, no reason to think that Darwin has questioned fixity at this time.

30. Darwin Papers, vol. 42.

31. C. Darwin, *Journal of Researches* (London: Henry Colburn, 1839), p. 208 (hereafter cited as *JR*); and C. Darwin, *Geological Observations on the volcanic islands and parts of South America visited during the voyage of H.M.S. 'Beagle',* 3d ed. (New York: Appleton, 1897), pp. 345–49.

32. These reflections follow the February 1835 note; they refer to W. Buckland, "Appendix. On the occurrence of the remains of Elephants, and other Quadrupeds, in the cliffs of frozen mud, in Escholtz Bay, within Beering's Strait, and in other distant parts of the shores of the Arctic seas," in R. W. Beechey, *Narrative of a Voyage to the Pacific and Beering's Strait,* 2 vols. (London: H. Colburn & R. Bentley, 1831) 2: 331–56.

33. C. Lyell, *Principles,* 1: 140.

34. C. Lyell, *Principles,* 2: 32–33.

35. C. Lyell, *Principles,* 2: 128–30.

36. C. Lyell, *Principles,* 2: 128–30.

37. C. Lyell, *Principles,* 2: 128–30.

38. C. Lyell, *Principles,* 2: 128–30.

39. See the informative study J. H. Ashworth, "Charles Darwin as a Student in Edinburgh, 1825–27," *Proceedings of the Royal Society of Edinburgh* 55 (1934–35): 97–113, where parts of Darwin's notebooks are reproduced in facsimile. A full transcription of the passages on the ova of *Flustra* is now available in P. H. Barrett, ed., *The Collected Papers of Charles Darwin,* 2 vols. (Chicago and London: University of Chicago Press, 1977), 2: 285–93. For another, more general discussion almost contemporary with Darwin's and Grant's work, see J. V. Thompson, *Zoological Researches and Illustrations* (Cork: King & Ridings, 1828–34), memoir 5. This memoir, on "Polyzoa," was first printed in 1830. A facsimile reprint of Thompson's *Researches,* with an introduction by A. Wheeler, has been published by the Society for the Bibliography of Natural History (London, 1968). L.

Hyman surveys the literature (from 1558 on) on "Zoophytes" in her treatise *The Invertebrates* (New York, London, and Toronto: McGraw-Hill), 5: 275–78 and 501–15.

40. R. E. Grant, "Observations on the Structure and Nature of the Flustrae," *Edinburgh New Philosophical Journal* 3 (1827): 107–18 and 337–42. Two of Darwin's letters to Henslow are instructive here. For they show that Darwin was himself very consciously dividing his observing, collecting, and speculating between his "*geological* section!" and his "second *section* Zoology" (the allusion is of course to divisions of labor in the new British Association for the Advancement of Science). And they show, too, that his lifelong preoccupation with "propagation" and "gemmules," later to surface publicly in the theory of pangenesis, traces directly back to Grant, Edinburgh, and *Flustrae.* See his discussion of "propagation" and of "the gemmule" in certain corallines in the July 24, 1834 letter to Henslow; for the geological and zoological "sections," see the letter of March 1834 (Barlow, *Darwin and Henslow,* pp. 83–98). On the tree cuttings at Chiloe, see Darwin Papers, 31 (i): 266–67.

41. Darwin Papers, vol. 33.

42. C. Darwin, "A Sketch of the Deposits Containing Extinct Mammalia in the Neighbourhood of the Plata"; this paper was presented at the meeting of May 3, 1837, then published in *Proceedings of the Geological Society of London* 2 (1838): 542–44, and is available now in Barrett, *Collected Papers,* I: 44–45. Note that, despite its title, it discusses the southern plains as far south as Port St. Julian.

43. For the text, commentary, and dating, see S. Herbert, ed., "The Red Notebook of Charles Darwin," *Bulletin of the British Museum (Natural History). Historical Series* 7 (1980) (The Red Notebook is hereafter cited as RN; pages cited refer to manuscript page numbers.)

44. RN 132; Kohn, "Theories," pp. 79–80.

45. *JR* 211–12.

46. *JR* 261–62.

47. C. Lyell, *Principles,* 2: 124–26.

48. C. Lyell, *Principles,* 2: 127.

49. C. Lyell, *Principles,* 2: 127.

50. C. Lyell, *Principles,* 2: 128.

51. C. Lyell, *Principles,* 2: 176–77.

52. C. Lyell, *Principles,* 2: 175–77.

53. For expert views of Lyell's innovation, see J. D. Hooker, "Presidential Address to Section E (Geography)," in *Report of the Fifty-first Meeting of the British Association for the Advancement of Science* (London, 1882), pp. 727–38; the Introductory Essay in J. D. Hooker's *Flora Novae Zelandiae, Part I: Flowering Plants* (London, 1852), and Asa Gray's review of it in *American Journal of Science and Arts,* 2d ser. 17 (1854): 241–52 and 334–50; A. R. Wallace, "On the Law Which has Regulated the Introduction of New Species," *Annals and Magazine of Natural History,* 16 (1855): 184–96; and the correspondence between Lyell and Forbes printed in K. Lyell, *Life, Letters, and Journals,* 2: 106–13.

54. See, further, M. J. S. Hodge, "Origins and Species," Ph.D. dissertation, Harvard University, 1970.

55. See, for example, Herbert, "Darwin's Theory of Transmutation: Part I," p. 235.

56. C. Lyell to Charles Lyell, Sr., February 7, 1829, in K. Lyell, *Life, Letters, and Journals,* 1: 245–46.

57. C. Lyell, *Principles of Geology,* 9th ed. (London: John Murray, 1854), pp. 702–3.

58. A. De Candolle, "Geographie botanique," in F. G. Levrault, ed., *Dictionnaire des sciences naturelles* 18 (1820): 416. For a recent discussion of De Candolle, see G. Nelson, "From De Candolle to Croizat: Comments on the History of Biogeography," *Journal of the History of Biology* 11 (1978): 269–305.

59. De Candolle, "Geographie botanique," p. 419.

60. See W. F. Cannon, "Charles Lyell, Radical Actualism, and Theory," *British Journal for the History of Science* 9 (1967): 104–20, and "The Whewell-Darwin controversy," *Journal of the Geological Society of London* 132 (1976): 377–88.

61. For a fuller discussion of these aspects of Humboldt's biogeography see Hodge, "Origins and Species."

62. C. Lyell, *Principles,* 1: 55–75, and the entry "Geology, geognosy" in the Glossary appended to the third volume.

63. C. Lyell, *Principles of Geology,* 9th ed., pp. 146–7.

64. See Herbert, "Logic of Darwin's Discovery," "Darwin's Theory of Transmutation: Part I," and "Darwin's Theory of Transmutation: Part II"; Richardson, "Darwin's Ideas on Transmutation"; and Kohn, "Theories." In addition see M. Kottler, "Charles Darwin's Biological Species Concept and Theory of Geographic Speciation: The Transmutation Notebooks," *Annals of Science* 35 (1978): 275–97; F. J. Sulloway, "Geographic Isolation in Darwin's Thinking: The Vicissitudes of a Crucial Idea," *Studies in History of Biology* 3 (Baltimore: Johns Hopkins University Press, 1979), pp. 23–65; and Herbert's introduction to Red Notebook in "Red Notebook."

65. For a detailed discussion of the difficulties of dating this entry in the personal pocket diary, see R. Colp, "'I was born a naturalist': Charles Darwin's 1838 Notes about Himself," *Journal of the History of Medicine and Allied Sciences* 35 (1980): 8–39. A copy of the diary was published in G. de Beer, ed., "Darwin's Journal," *Bulletin of British Museum (Natural History). Historical Series* 2 (1959): 1–21.

66. Darwin's reading and annotating of this edition of Lyell have been much discussed over the years, and most recently in Kohn, "Theories," p. 155 and Herbert, "Darwin's Theory of Transmutation: Part I," p. 257–58, and "Darwin's Theory of Transmutation: Part II," pp. 201–2. Two things make it seem most probable that most of the annotations were made in late summer 1838 (but before reading Malthus on September 28): the overlap between the content of the annotations themselves and many notes, made at that time, in Darwin's notebooks C and D, and Darwin's entry in his reading notebook relating to this edition of Lyell. The reading notebooks have been published in P. J. Vorzimmer, "The Darwin Reading Notebooks (1838–1860)," *Journal of the History of Biology* 10 (1977): 107–53 (see p. 120). All references to and quotations of the B, C, D, and E notebooks here are based on comparing the two fullest transcriptions with one another: G. de Beer, ed., "Darwin's Notebooks on Transmutation of Species," *Bulletin of the British Museum (Natural History). Historical Series* 2 (1960): 23–183, as supplemented by G. de Beer et al., eds., "Darwin's Notebooks on Transmutation of Species. Pages Excised by Darwin," *Bulletin of the British Museum (Natural History). Historical Series* 3 (1967): 129–76; and P. H. Barrett, ed., *Computerized Printout: Darwin's Transmutation Notebooks—B, C, D, E* (Lansing: Michigan State University, 1972). Where the passage is included there, I have taken as authoritative Barrett's text, as given in "Extracts from the B, C, D, E Transmutation Notebooks," in Gruber and Barrett, *Darwin on Man,* pp. 439–60. Judgments as to which entries in Notebook B were made as later insertions by Darwin are based on an examination of the notebook itself, in the Darwin Papers, vol. 121.

67. L. Wilson, *Charles Lyell, the Years to 1841: The Revolution in Geology* (New Haven: Yale University Press, 1970); chap. 8; and Herbert, "Darwin's Theory of Transmutation: Part I," pp. 241–45.

68. Kohn, "Theories," p. 83; Gruber, "Psychological Study of Scientific Creativity," pp. 135–43.

69. Kohn, "Theories," pp. 83–88.

70. Cf. E 54 for more on the "laws of life." See section 10 of this essay for Darwin's preoccupation with this phrase.

71. The recent articles, cited above, of Herbert, "Darwin's Theory of Transmutation: Part I," and "Darwin's Theory of Transmutation, Part II"; Kottler, "Transmutation Notebooks"; Kohn, "Theories," and "Charles Darwin's Path"; and Sulloway, "Geographic Isolation," all endorse some such dating.

72. De Beer, "Darwin's Journal" (i.e., his personal pocket diary), p. 7.

73. Barlow, *Darwin and Henslow,* pp. 123–31; and Herbert, "Darwin's Theory of Transformation: Part I," pp. 245–49.

74. See the published versions of the papers as reprinted in Barrett, *Collected Papers,* 1: 38–45.

75. See Herbert's introduction to the Red Notebook in "Red Notebook."

76. Kohn, "Theories," pp. 73–75. Kohn provides a detailed interpretation of Darwin's consideration of the various transmutation possibilities.

77. C. Lyell, *Principles,* 2: 118–19 and 89.

78. For a more precise analysis, see Kohn, "Theories," pp. 73–81.

79. C. Lyell, *Principles,* 1: 128–30, and 2: 150–54.

80. Herbert, "Logic of Darwin's Discovery"—see especially pp. 1–45—provides the fullest documentation of these preoccupations.

81. Barlow, *Diary of the Voyage,* pp. 394–400 and 409–13.

82. Herbert, "Darwin's Theory of Transmutation: Part I," p. 236, where Gautrey's dating is cited.

83. The "Reflections" are in vol. 42 of the Darwin Papers.

84. Note that the skunk (under the name Zorilla) and its migrations will show up again, at RN 128.

85. Barlow, *Diary of the Voyage,* p. 236.

86. C. Lyell, *Principles,* 1: 140.

87. Darwin to Henslow, July 12, 1835, in *Darwin and Henslow,* pp. 109–11.

88. *Darwin and Henslow,* pp. 109–11.

89. Barlow, *Diary of the Voyage,* p. 295. See later additions made to notes for 1834, in Darwin Papers, 31: 278r and following pages.

90. See Barlow, *Charles Darwin and the Voyage of the Beagle,* p. 247, and Barlow, *Diary of the Voyage,* p. 337.

91. See Darwin Papers, 31: 328r, 340r, 341r, and 342r.

92. Barlow, *Diary of the Voyage,* pp. 372 and 383.

93. See C. Lyell, *Principles,* 1: 159–60.

94. Barlow, *Diary of the Voyage,* p. 383.

95. N. Barlow, ed., "Darwin's Ornithological Notes," *Bulletin of the British Museum (Natural History). Historical Series* 2 (1963): 203–78; Herbert, "Darwin's Theory of Transformation: Part I," pp. 236–40. For the Burchell references, see MS p. 82. Barlow's discussion of the *Rhea* entries is at pp. 273–76.

96. *ON* 71 (the MS page numbers are cited).

97. *ON* 72.

98. *ON* 42.

99. *ON* 73.

100. C. Lyell, *Principles,* 1: 126–27; *ON* 73–74; and Wallace notebook (1854), at Linnaean Society, London, pp. 45–46. For dating and partial quotation, see H. L. McKinney, *Wallace and Natural Selection* (New Haven and London: Yale University Press, 1972) pp. 32–37.

101. For the Falkland fox, see Darwin Papers, 31: 237r.

102. *ON* 74. Note: it is absolutely essential to distinguish, in Darwin's thinking from this point on, between the transmutations of species he thinks occasioned by the original mainland-to-island migrations and any further divergences occasioned by subsequent island-to-island migration. His initial concern is with the former, and that remains the primary issue; for island-to-island divergences (whether into distinct varieties as with the mockingbirds in 1836 or into distinct species as with those same birds when classified by Gould in 1837) are only of importance because of Darwin's prior interest in those prior transmutations; it is those transmutations of colonists from the mainland that have allowed many islands, some in archipelagoes some not, to have peculiar, indigenous species. Thus, it is telling that his conjecture (at B 6–7) about "a pair" of dogs or cats on a "fresh island" increasing slowly because of heavy predation ("from many enemies"), clearly echoes the discussion (in *JR* 248–49) of the distinctive rabbits, descended from introduced stock, on the Falkland Islands: "The first few pairs moreover had here to contend against pre-existing enemies, in the fox, and some large hawks." What Gould's classification of the mockingbirds will do for Darwin, in 1837, is to give him even better evidence for mainland-to-island transmutations than he had before, by establishing that island-to-island migrations had occasioned not merely varietal divergences but specific divergences. It would, therefore, be quite mistaken to think that Gould's classifying results first convinced Darwin that he needed transmutation in order to explain the biogeography of islands—or mainlands. See also my remarks in the closing section of this essay.

103. Barlow, *Diary of the Voyage,* pp. 394–413.

104. *ON* 80–85.

105. Kohn, "Theories," pp. 77 and 80. D. Ospovat, "Perfect Adaptation and Teleological Explanation," *Studies in History of Biology* 2 (Baltimore: Johns Hopkins University Press, 1978), pp. 33–56, and "Darwin after Malthus," *Journal of the History of Biology* 12 (1979): 211–30. For the quoted phrase, see G. de Beer, ed., "Darwin's Notebooks on Transmutation of Species," p. 167 (ms. p. E 72).

106. Cannon, "Darwin's Achievement."

107. On degeneration, see Kohn, "Theories," pp. 72-81.
108. For the sources in Lyell on Lamarck and in Erasmus Darwin, see Kohn, "Theories," pp. 83-85 and 109-11.
109. M. Ruse, "Darwin and Herschel," *Studies in History and Philosophy of Science* 9 (1978): 328-31; S. S. Schweber, "The Origin of the *Origin* Revisited," *Journal of the History of Biology* 10 (1977): 229-316; and E. Manier, *The Young Darwin and His Cultural Circle* (Dordrecht and Boston: D. Reidel, 1978).
110. See Gruber, "Psychological Study of Scientific Creativity," p. 127, for the diagrams. See Kohn "Theories," pp. 109-13 for the corrections.
111. *JR* 568; Barrett, *Collected Papers,* 1: 45 and 48.
112. Herbert, "Darwin's Theory of Transformation: Part I."
113. Herbert's article, just cited, gives an account of the interactions with Gould.
114. *ON* 78. See C. Lyell, *Principles,* 2: 38; and *JR* 475-78.
115. Kohn, "Theories," p. 156; and F. N. Egerton, "Darwin's Early Reading of Lamarck," *Isis* 67 (1976): 452-56.
116. C. Lyell, *Principles,* 2: 3, and Bartholomew, "Lyell and Evolution."
117. C. Lyell, *Principles,* 2: 3-12.
118. C. Lyell, *Principles,* 2: 12-17.
119. C. Lyell, *Principles,* 2: 60-65.
120. C. Lyell, *Principles,* 2: 3. See, further, M. J. S. Rudwick, "Charles Lyell's Dream of a Statistical Paleontology," *Paleontology* 21 (1978): 225-44.
121. C. Lyell, *Principles,* 2: 2.
122. C. Lyell, *Principles,* 2: 18-20.
123. C. Lyell, *Principles,* 2: 23.
124. C. Lyell, *Principles,* 2: 24-28.
125. C. Lyell, *Principles,* 2: 24-28.
126. Gruber, "Psychological Study of Scientific Creativity"; and Kohn, "Theories."
127. E. Darwin, *Zoonomia, or the Laws of Organic Life,* 2 vols. (Dublin: Byrne & Jones, 1794) 1: 569.
128. E. Darwin, *Zoonomia,* 1: 563.
129. Kohn, "Theories," pp. 83-86.
130. *JR* 261-62.
131. *JR* 261-62. See also Kohn, "Theories," pp. 85-88.
132. Gruber, "Psychological Study of Scientific Creativity," chap. 7. And Kohn, "Theories," pp. 109-13.
133. C.G. Ehrenberg, "On the Origin of Organic Matter from Simple Perceptible Matter and on Organic Molecules and Atoms," in R. Taylor, ed., *Scientific Memoirs* (London, 1837). See also *Edinburgh New Philosophical Journal* (1832), pp. 319-28. See, further, Kohn, "Theories," p. 161.
134. See, further, Kohn, "Theories," pp. 109-13.
135. For this division of the 1842 Sketch, see M. J. S. Hodge, "The Structure and Strategy of Darwin's 'Long Argument,'" *British Journal for the History of Science* 10 (1977): 237-46.
136. For a much fuller discussion, see Kohn, "Theories," pp. 81-100. Kohn's whole discussion of Notebook B should be consulted in connection with the explication given here.
137. For the Quinarian and other such systems, see E. Streseman, *Ornithology from Aristotle to the Present* (Cambridge, Mass.: Harvard University Press, 1975), and M. P. Winsor, *Starfish, Jellyfish, and the Order of Life* (New Haven: Yale University Press, 1976).
138. See de Beer, "Darwin's Journal."
139. See Kohn, "Theories," pp. 113-40.
140. R. Hooykaas, *The Principle of Uniformity in Geology, Biology, and Theology* (Leiden: E. J. Brill, 1963), pp. 100-106.
141. E. Mayr, *The Growth of Biological Thought: Diversity, Evolution, and Inheritance* (Cambridge, Mass., and London: Harvard University Press, 1982), pp. 391-92 and 399-400.
142. For a sampling of these two sets of issues, see E. Mayr, *Evolution and the Diversity*

*of Life: Selected Essays* (Cambridge, Mass., and London: Harvard University Press, 1976), and S. J. Gould, *Ever Since Darwin: Reflections in Natural History* (Harmondsworth and New York: Penguin Books, 1980).

143. For Darwin's "metaphysics," see his M and N notebooks, printed in full in Gruber and Barrett, *Darwin on Man.*

144. See S. S. Schweber, "The Origin of The *Origin,*" and "The Genesis of Natural Selection—1938: Some Further Insights," *Bioscience* 28 (1978): 321-26.

145. See Cannon, "Darwin's Achievement."

146. See, for example, J. R. Moore, *The Post-Darwinian Controversies* (Cambridge: Cambridge University Press, 1979).

147. T. S. Kuhn, *The Structure of Scientific Revolutions* (Chicago: University of Chicago Press, 1962); I. Lakatos, "Falsification and the Methodology of Scientific Research Programmes," in I. Lakatos and A. Musgrave, eds., *Criticism and the Growth of Knowledge* (Cambridge: Cambridge University Press, 1970), pp. 91-195; and L. Laudan, *Progress and Its Problems* (Berkeley and Los Angeles: University of California Press, 1977).

148. I. Scheffler, "Vision and Revolution: A Postscript on Kuhn," *Philosophy of Science* 39 (1972): 366-74.

149. Nineteen eighty-two sees the appearance of two highly informative and insightful articles by Frank Sulloway, both concerning Darwin in the 1835-37 period, in volume 15 in the *Journal of History of Biology:* "Darwin and His Finches: The Evolution of a Legend" and "Darwin's Conversion: The *Beagle* Voyage and Its Aftermath." A main thesis of the first paper—that the finches were of little importance for Darwin's theorizing about the origin of species in these years—fits fairly well with the conclusions reached in the present essay; but a leading proposal in the second—that John Gould's decisions about the mockingbirds brought about Darwin's first serious commitment to the thesis that new species arise by transmutation of older ones—obviously does not. Detailed pursuit of such disagreements always provides an opportunity to throw into sharper relief several worthwhile and wide-ranging historiographical and exegetical issues; I examine such issues in a discussion paper now in preparation.

# Experimental Biology in Transition: Harrison's Embryology, 1895–1910

*Jane Maienschein**

## "The Experimental Ideal"

Idealism gave way to materialism in biology around 1900, Garland Allen has told us in his widely read textbook on the history of twentieth-century biology.[1] Young biologists rejected the idealism, phylogenetic concerns, and descriptive methods of their predecessors and turned to the physical sciences as a model, according to this view. This led to their endorsement of experimentation. "No biologist could long resist the temptation of such a promising method," William Coleman has proclaimed in his equally well-known textbook on nineteenth-century biology.[2] Especially in the United States, the cause of experimental analysis, as exemplified by Wilhelm Roux's "Entwickelungsmechanik," gained "almost immediate support," Allen has maintained.[3] The result, Allen and Coleman have agreed, was a revolt in biology away from traditional descriptive morphological science to experimentation as the ideal.[4]

This picture of a generation of young rebels casting off the shackles of worn-out speculation and observation and embracing a brave new scientific method is appealing indeed. Unfortunately, the picture depends on boldly overdrawn distinctions, which I think that neither Allen nor Coleman would now support without further clarification.[5] In contrast, I therefore propose, by sketching the portrait of one central figure, to suggest an alternative picture of early twentieth-century biology that demonstrates evolving assumptions about, rather than revolutionary changes in, what proper methodology in biological investigation should be.

Specifically, I find that Ross Harrison's work in experimental embryology illustrates important methodological patterns and changes that exemplify developments in experimental biology generally. Above all,

*Research for this paper was supported by a National Science Foundation grant, SOC-7912542. I also wish to thank William Coleman for his insightful and valuable suggestions. Department of Philosophy and Humanities, Arizona State University. 

Harrison's work reveals that we must remain very clear about exactly what we mean by such terms as experimentation. Experimental science has obviously evolved since Bacon's and Descartes's time and has always meant different things to different people. Equally, experimentation varied for such noted "experimentalists" as Thomas Hunt Morgan or Frank Rattray Lillie, for example, and Harrison's case demonstrates that experimentation could develop in different ways even within one person's research.[6]

To help clarify discussion, I feel that it is essential to avoid adopting any single experimental standard but necessary to attempt to deal with the historical complexities. I suggest that there actually exists a continuum of experimentalisms, ranging from utilizing simple manipulative techniques to endorsing fully developed experimental hypothesis testing within a research program.[7] Harrison's evolving style, which exemplifies that of the "new experimenters," reveals such experimental options. In his earliest work, he did not seek causal explanations of embryonic development, as Roux did. Rather he wanted to delineate the developmental processes. It is thus important to recognize that there can be different types of experimentation and different problems or goals for which those different experimentations may be appropriate.

All levels of biological experimentation involve exercising control over organisms, but the nature and purpose of that control varies significantly. One can use a range of manipulative techniques beginning with simple disruption of normal conditions to elaborate construction of complex abnormal situations. The researcher can use the techniques to ask simply, "What will the organism do under such-and-such conditions?" Or the techniques can be applied to test hypotheses and thereby to address specific well-formulated questions with an experimental approach. If the hypothesis in question is embedded within a coherent research program, the experimental approach reaches its most complex form. In discussing whether early twentieth-century biology became experimental, we must be particularly alert to what we mean by that term.

The evolution of Harrison's work, as revealed through his published writings, demonstrates the way in which he came to endorse consciously an experimental approach within a research program and why he felt it important to do so.[8] Beginning with relatively simple experimental manipulations to obtain new data in his early work, he soon decided that his results were not satisfactory; he therefore articulated an experimental and theoretical context for his manipulations. His emerging endorsement of an experimental approach exemplifies what Coleman, Allen, and others have considered the heart of early twentieth-century biology. The facts in this one case will therefore reveal an alternative to Allen's and Coleman's understandings of the reasons for the endorsement of experimentation.

## Ross Harrison's Medical Orientation

When Harrison began his early work, he confidently reported the results of his research as conclusive contributions to understanding development. In 1895, for example, his first major paper, which was his doctoral thesis, appeared in the *Archiv für Mikroscopische Anatomie*. Entitled "Die Entwicklung der unpaaren u. paarigen Flossen der Teleostier," it immediately followed a paper by Harrison's German mentor Moritz Nussbaum.[9] Harrison's concerns and his style paralleled Nussbaum's rather closely, which should not be surprising since the research was carried out at Nussbaum's suggestion and in his laboratory at Bonn. Harrison clearly had become a successful user of experimental manipulations. But he was operating within the context of traditional German medical school studies, with the goals of answering questions about anatomy and embryology in order to have a full description of each—for ultimately practical reasons. There is absolutely no evidence that he had given "almost immediate" support to Roux's experimental program of "Entwickelungsmechanik." Harrison did not seek to establish causal connections in development, as Roux did, nor did he endorse the "experimental ideal" of hypothesis testing in a more modern sense.

Harrison's background at the Johns Hopkins University, which historians of American science have liked to emphasize, played a less direct role in shaping the problems or methodology in his early dissertation work than it did later.[10] His graduate study and his undergraduate work at the same medically-oriented new American university did convince Harrison to go to Germany to study, but this was not unusual since most of his successful contemporaries also went.[11] While not denying that William Keith Brooks's and H. Newell Martin's biological influence on Harrison may have been significant in the long run, I think it did not shape his work as much as did his decision to study medicine in Germany. In Germany, Harrison settled in an anatomical institute in Bonn and completed his degree in medicine, which others of his contemporaries who became leading figures in biology did not. Why? And what influence did Harrison's decision to work on anatomical and embryological problems in a German medical context have on his later commitments?

In answering this question of Why?, we encounter the familiar and difficult problem of establishing influence. In the face of negligible clues from archival sources or published material, what can we conclude about influence? Did Harrison choose a medical context purposefully or by chance? Apparently, there was a certain measure of the latter, according to the only apparent extant source.[12] Chance placed him with Nussbaum at least, but interest in the problems of fin development in teleosts made him visit Bonn and Nussbaum in the first place to seek advice.

Perhaps Harrison felt comfortable with the fortuitous choice of a medical setting at Bonn because of his own prior medical orientation; according to reports, he had almost chosen medicine for his original course of study in America.[13] Neither Bonn nor Nussbaum was obscure scientifically, so Harrison's choice was quite reasonable. Bonn had become a famous medical center when Johannes Müller held the chairs of anatomy and physiology there, followed briefly by Hermann von Helmholtz. The subsequent tenure of Max Schultze as professor of anatomy and Ernst Pflüger as professor of physiology continued the tradition of excellence in medicine. Nussbaum studied under Schultze and Pflüger, then stayed in Bonn, progressing slowly through the ranks in physiology until 1907, when he was given a personal chair as Ordentlicher Professor in Biology and Histology. Thus Harrison's choice of Bonn and Nussbaum may not have been fully premeditated, but it was hardly an unlikely choice—given Harrison's medical orientation.[14] His choice of teleost fin development is more difficult to identify, though it is unlikely that his interest grew from his summer in 1890 at Woods Hole, as so many of his contemporaries' interests did. There he had worked on early oyster development—not a likely stimulus for fin investigation![15]

Whatever the reasons, it is clear that Harrison chose to pursue a medical degree self-consciously. He purposely did not spend much time at the Naples Zoological Station while in Europe, as his friend Morgan had done, and he similarly did not seek a mentor with evolutionary interests, as others of his colleagues from Johns Hopkins had.[16]

The second question remains. What difference did his choice of a basically medical context make to his later work? How did Harrison's career differ because he sought a medical degree rather than continuation of research for a doctorate in zoology while in Germany? Here again, the difficult problem of establishing influence arises. But this is a different type of influence, an assessment of which must begin by considering whether medicine and biology were substantially enough differentiated that it is legitimate to study the influence of one on the other.

There exists no clear dichotomy of medicine as absolutely distinct from biology. In fact, the two are now—and were then—clearly related and overlapping. Yet, despite the difficuties in identifying some work as either biological or medical, the fact remains that by 1890 many people *were* eager to distinguish the two fields.[17] As Coleman has stated,

> Sometime during the middle third of the eighteenth century those interested in the phenomena of life began to isolate and examine special problems for consideration and, knowingly or otherwise, to devise or articulate special techniques and viewpoints for prosecuting their examination. This process raced on undiminished throughout the nineteenth century. Its ultimate effect was to create a body of men who were recognizably biologists and whose subject, embracing a multitude of specialties, was biology. The creation of biology as a recognized

discipline thus followed with only brief delay upon the determination of the legitimate subject matter of the science.[18]

Also, the fact that the Johns Hopkins University had a medical school, with a department of anatomy, and a separate graduate school with a completely autonomous department of zoology, attests that there existed at least perceived differences between zoology (or biology) and medicine.

There are obvious grounds for distinction stemming from the fact that medicine deals with disease, focuses on diagnoses, and seeks therapy by way of clinical practice. As a result, the two areas were primarily distinguished by subject matter, as Coleman has suggested, and also by the type of result sought rather than by methodologies.[19] To help clarify any real distinctions, consider a range of problems from the very practical and precisely defined, i.e., concern with correcting specific structural or functional defects in humans, to the very theoretical and general, i.e., seeking careful descriptions and explanations for phenomena of life in general. In 1890, traditional medicine was identified with the former end of the range and the "new experimental biology" with the latter. Most biological, medical, and biomedical work belonged between the extremes, of course, but researchers did generally identify with one extreme or the other and therefore perceived themselves as either biologists or medical workers.

In 1890 there was an important constraint on biology, which restricted the type of explanations considered desirable or acceptable. From Darwin and Haeckel to Roux and the other advocates of experimentation and causal explanation, all stressed evolution. Biological results must not only conform to evolution theory, but it was agreed that evolution theory could help lead to desired explanations.[20] Thus even though Morgan won a Nobel Prize in Physiology and Medicine and even though much of this work lay in a middle ground between medicine and biology, such men as Edmund Beecher Wilson, Edwin Grant Conklin, Morgan, Lillie, Charles Manning Child, William Morton Wheeler, George Howard Parker, and Herbert Spencer Jennings all identified with evolution in some way and considered themselves biologists.

In significant contrast, Harrison both rejected evolutionary constraints on his work and sought a medical degree. He did not deny that life had evolved, but he felt the whole study of evolution too speculative to be studied scientifically.[21] Evolution should not be either a regulative or directive constraint on biological research, according to Harrison. This difference from his contemporaries who also became leaders in biology was basic to Harrison's ability to forge an experimental embryological program, a program that by 1910 he considered *biological* despite its absence of evolutionary constraints.[22] And this biological program represented a different biology than that of the 1890s.

His fellow students at Johns Hopkins and elsewhere had, like

Harrison, begun their research with describing embryological development, but all moved into other specialty areas of zoology. Because of Harrison's deep-rooted concern with embryology as at least partly a *medical* problem in the traditional anatomical sense, emphasizing careful description of each stage of development against a background hope for practical medical applications, Harrison's emerging *biological* concern with seeking explanations of developmental processes differed from his friends' work. Harrison came to be considered both a medical and a biological researcher by others, though by 1910 he considered himself more a zoologist and his program of experimental embryology an essentially biological program.[23] His description of zoology in 1914, at the dedication of the new Osborn Memorial Laboratories for Biology, at Yale, reveals the differences but also the close alliance he saw between biology and medicine:

Biology embraces that whole group of sciences which deal with the living world. They had their origin in the practical needs of medicine and in the study of natural history, which was the expression of the desire of man to know something about his fellow beings, to name them and to classify them. This group of sciences is complex, and, being a growth in adaptation to various practical exigencies, its subdivisions are not altogether logical. In the broadest sense, zoölogy, which has to do with animal life, is half of biology, botany the other half. According as animals are studied with respect to their form and structure, their activities, or their condition in disease, we distinguish morphology or anatomy, physiology, and pathology; but these subjects, which constitute the immediate foundation of modern medicine, are studied so intensively with reference to man and his more closely related brethren that special laboratories are provided for their accommodation, and those who devote themselves to them would no doubt resent their being classed as mere subdivisions of zoölogy. There is, however, an immense field including the structure, functions, life histories and relationships of animals that is scarcely touched upon by the above-named subjects. This is zoölogy in the more restricted sense, and, together with the study of plants, is that which is usually known in the college curriculum as biology. The problems of zoölogy, however, are to a considerable extent similar to and even identical with the problems of anatomy and physiology. In other words, zoölogy speaks the language of anatomy and physiology, but besides this it has to have command of the dialects of those less organized regions where such subjects as animal behavior, geographical distributions, ecology or the adaptation of organisms to their environment, evolution, and heredity are cultivated.[24]

Zoology, in other words, was simply at one end of a continuum with medicine at the other. According to his biographer John S. Nicholas, Harrison "sometimes whimsically referred to medicine as applied biology."[25] The type of work he valued in embryology originally grew out of what was considered a medical environment; then it came to be considered biological. But the work actually lay between the traditional disciplines of biology and medicine, illustrating the evolving nature of disciplines as well as of experimental methodologies. Harrison's biology by 1910 was a new biology, different from that of the 1890s.

Harrison's career and his resulting embryological program depended in an essential way on his close identification of biology and medicine and on his choice to ignore evolution as an explanatory or directive factor in research, a commitment which likely influenced him to study nerve and muscle development and to go to an anatomical institute in Germany. His commitment was reinforced there by his participation in a medical, nonevolutionary environment as it would not have been had he studied instead at the Naples Zoological Station. The remainder of this paper illustrates Harrison's move from a more traditional descriptive approach that he considered medical to a more biological approach at Yale. It explains how Harrison's changing assumptions about appropriate experimental methodology and the related shift in the problems that he sought to address were essential to that move.

## Harrison's Experimentation

Concerned at Bonn with tracing the role of cells rather than germ layers in development, Harrison described anatomical and successive embryonic stages with impressive competence.[26] Yet he felt a weakness because he could not give details of the development of nerves, a subject he found particularly interesting. Unfortunately, he wrote, "about the development of nerves in the unpaired fins, I could indicate nothing of significance. With the help of the usual embryological methods the particulars of development are not to be discovered."[27] These inadequate methods were based on experimental manipulations, designed to cut the organism into analyzable cell groups and to fix the action of normal development, and on description of the results of that intervention. They could not provide more data than observation of normal development allowed.

Although Harrison and Nussbaum wanted to understand the dynamic processes of development, their experimental methods did not involve the controlled hypothesis testing characteristic of more sophisticated experimentation and were not, like Roux's work, oriented toward achieving causal analysis of the processes. As if description of sequential stages of development would reveal the significant connections, Harrison, like the contemporary group of "cell counters" doing cell-lineage studies in Woods Hole in the 1890s, depended on observation and on description of embryonic stages to address a question of description: What happens during ontogeny? This work was experimental only in the most basic sense of using simple manipulations to produce and fix disrupted and hence necessarily abnormal specimens and thereby to gain control over observations. The method differed much more from the experimental approach Harrison endorsed soon afterwards than from the traditional manipulations used throughout the nineteenth century by German medical researchers in anatomy as well as in physiology. His use of particular cutting, fixing, and staining techniques represented typical descriptive medical research. Again we see that Harrison fit very well within his environment at the anatomical institute at Bonn. And his work did yield productive results for advancing anatomical description of stages of organic development. But Harrison had not lent support, either immediate or otherwise, to Roux's program of "Entwickelungsmechanik." There is, in fact, no evidence that he had at that time any interest in Roux's suggestions.[28]

Harrison's introduction to Gustav Born's special techniques for transplanting parts of embryos onto other embryos provided the basis for more analytical ways to address questions about nerve development, problems the old methods could not touch.[29] In a paper of 1898, in which he used Born's techniques, Harrison wrote that "in the method of grafting we have a means of experimentation for which no substitute is offered. Born's discovery that certain amphibian embryos lend themselves with readiness to such operations is of especial importance in that it renders the method applicable to the study of developmental problems."[30] Born's method allowed the researcher to graft togther parts from two differently pigmented species. The hybrid continued to develop as one organism with two differently colored areas, which could thereafter easily be followed throughout development. "By varying the region in which the parts are stuck together," Harrison enthused, "it thus becomes possible to trace out the mode of growth of individual structures or organs."[31]

Even though Harrison's presentation remained thoroughly formal, the reader can sense a certain excitement in his realization of the importance of Born's technique. Yet in 1898 Harrison used the remakable technique simply to describe in more detail than had previously been possible how cells act in development and regeneration (in this case the cells of frog tails). He saw the promise of the method as being the production of additional data for an anatomical understanding of organisms' development, not a tool for hypothesis testing. His grafting experiments provided data that could test the accepted view of, for example, "the individual constituents of the tail."[32] But such tests were really descriptive tests, answering "Is it true that what they say is really what happens?" Careful observation using Born's techniques, therefore, provided new data, not new explanations. The reader gains no sense from this paper of why Harrison—or anyone else—would want to know these facts except that they contributed small pieces to a descriptive picture of embryonic development. Nowhere did Harrison reveal a broader perspective or larger question that might have motivated his research. The problems for which he used his newfound experimental techniques still concerned careful anatomical and embryological description.

In retrospect, some historians have been tempted to regard this as less interesting than fuller use of an experimental approach and to consider the rise of experimental hypothesis testing as more important for turn-of-the-century biology. Yet this careful use of experimental techniques to develop descriptive data also represents an essential element of the "new experimental biology." Harrison's medical work, like his contemporaries' cell-lineage studies, reveals a fundamental commitment to asking narrowly defined questions and to obtaining definitive results. At the time, his seemingly more traditional medical approach, still within a medical context that sought to utilize research results, was definitely legitimate and important.

Furthermore, Harrison's use of Born's grafting techniques, although initially unexciting to retrospective judgment, did move him in a different direction than had his earlier concern with describing normal developmental stages. He began to use abnormal or contrived conditions to learn more about the normal progression of cellular changes and changes in groups of cells, thus emphasizing both cells and dynamic change. Born's method provided a practicable means for analyzing development into distinguishable smaller pieces so that one could obtain apparently definitive results. It was a pivotal innovative experimental technique for manipulating development and thus going beyond normal conditions.

After his introduction to Born's techniques, Harrison continued along his descriptive morphological way, refining his observations and his experimental manipulations to address various embryological questions. Then, for the first time, in a paper of 1903 on lateral line sense organs in amphibians, he compared his experimental results with "control" or normal development, to the extent that he knew the latter, and he closed with a brief but noteworthy "Schluss"' section which discussed current general theories of development.[33] He referred to the views of Hans Driesch and Roux on the then much-debated issue of self-differentiation and concluded that the "Anlage" of the lateral line are already represented in the germ and develop without additional impulses being necessary. But, he concluded, it is the mutual effect of embryonic parts, at an early developmental stage, that determines that certain cells will become lateral line cells. Though no experimental data had illuminated the processes to date, Harrison added, he hoped that Hans Spemann's efforts would show the nature of the determining factors and the moment that differentiation becomes fixed.[34]

Awareness of Spemann's early studies, which he cited, and of the issues of self-differentiation and of concern with dependent differentiation within the totality of the organism refocused Harrison's research—or at least his published presentation of that research.[35] He began to report his results in terms of general theories of development, thus moving beyond his previous purely descriptive account that simply states "here it is." Instead he laid out his results and went on to conclude that they supported self-differentiation—or almost.

No doubt reinforcing Harrison's apparently expanded awareness of general issues was his attendance at an important scientific meeting held in April 1904. The Anatomische Gesellschaft convened in Jena to discuss a variety of current topics, of which nerve fiber development was one of the most central.[36] Harrison participated in a session with Albert von Kölliker and Oscar Schultze, who argued respectively that sheath or Schwann cells do and do not play a crucial generative role in fiber development.[37] Harrison, drawing on his own experiments in which he had removed the source of the Schwann cells, argued that they could not play a primary role, since nerve fibers developed even without their presence; they must instead play a secondary role, if any role at all.[38]

This meeting seems to have initiated Harrison into high-powered verbal exchange regarding his research subject. One is tempted to suggest that his realization that people are obstinate and that experimental "proof" is more difficult to provide than he had thought led him to seek a more definitive approach in his research. However this may be, his attendance undoubtedly did shape the way he oriented his studies after 1904. And, perhaps more significantly, the way he used experimentation and the way he reported his results began to change subtly as he saw the inadequacy of previous attempts. From 1904 to 1910 Harrison sought to discover what he had to do to make his work convincing—for one of his assumptions was that one's work should be convincing (meaning conclusive, compelling others to accept it as definitive).[39]

During this period Harrison concentrated on establishing beyond doubt the way in which nerves develop and the role of the nervous system in development. At first, he did not question his assumptions; thus he did not ask why one would want to know about the development of nerves or why one would focus on cells in nerve development. This made sense within the context of his essentially descriptive interest in body processes, getting results, and answering questions about what happens to make up the final anatomical product. For these ends, Harrison urged, success depended on proper use of experimentation. "It is clear," he wrote, ". . . that the facts are insufficient to determine even the comparatively simple relations between the nervous system and the developing musculature." Thus the only effective approach "is by direct experimentation, but in devising experiments for this purpose it is necessary to formulate clearly just what is to be determined, for it is obvious . . . that the nervous system may possibly exert its influence in a variety of ways."[40]

Therefore Harrison reported experiments that were executed specifically "in order to determine the effect of [one factor or another] on normal development." He tested, for example, the effect of suspension of muscular activity by treating the embryo with acetone chloroform. This allowed him to determine how the embryo developed without normal functional stimuli. Clearly, then, the experiment provided a way of testing the role of functional development by isolating that factor and eliminating it. The experimental results provided data not obtainable from normally developing organisms alone. Thus the experimentally induced pathological specimens yielded new data, data that Harrison felt were useful, although others disagreed with the validity of the data since they were "abnormally derived."[41]

At least this experimental method, designed to test a particular question, was potentially conclusive. Two years later Harriosn wrote:

> Prior to the year 1904 all attempts to solve these problems were based on observations made upon successive stages of normal embryos. When one compares the careful analyses of their observations, as given by various authors, one cannot but be convinced of the futility of trying by this method to satisfy everyone that any particular view is correct. The only hope of settling these problems definitely lies, therefore, in experimentation.[42]

General recognition of the value of an experimental approach—whatever that meant to the principals involved—was acknowledged by the establishment of the *Journal of Experimental Zoology* in 1904, with Harrison as managing editor.

In his experiments reported in 1904, Harrison showed that the muscles developed even after removing the spinal cord and hence the nerve tissue. Therefore "this experimental demonstration of the independence of the developing muscular tissue may be regarded as crucial evidence against the general correctness of the view . . . that the first development of the muscles takes place under the influence of the nervous system."[43] The results reinforced Harrison's belief that Roux's view of development as involving a separate organ-forming stage and functional development stage must be misleading; he saw the two stages as really overlapping and complexly interrelated.[44]

Perhaps because he had seen the different ways the same descriptive evidence could be used within the traditional medical context, as it had been at the 1904 meeting in Jena, in 1904 Harrison recognized the value of using experimental techniques to achieve what he considered to be "crucial evidence" for addressing a clearly formulated question. Yet he also recognized that he had so far only contributed to "one phase" of understanding the general problem of neural influence in development. At last—from the retrospective point of view—he understood that he had a general problem and that to achieve useful results it was productive to attack smaller phases one at a time. The product was the emergence of a specialized embryological research effort.

This is not meant to suggest that Harrison's effort was unique, for Spemann, in Germany, certainly provided a notable parallel, but is meant to emphasize that Harrison's specialization in experimental embryology was different from contemporary biologists' specializations in cell-lineage studies, cytology, and genetics.

As he pursued his studies, Harrison began to realize that his experimentation could succeed best in producing results when very narrowly defined questions were posed. Still stimulated by discussions of details at the Jena meeting in 1904, Harrison felt the additional stimulus of obtaining definitive results to address increasingly well-defined and specific questions. From 1906 to 1910 Harrison no longer asked general anatomical questions about structures or development of whole nerves or nervous systems. Instead he asked more specific questions about the nature and processes of nerve parts. In particular: What is the constitution of the nerve fiber and how does it grow from its point of origin to the periphery of the organism?

Beginning with this problem, Harrison by 1907 had formulated the basic elements of a research program. He knew what questions he wished to answer and what the current theoretical answers were. He knew that an experimental approach was essential to achieve productive results and that he should concentrate on cells as the fundamental developmental units for experimental analysis. But he also felt that he should keep in mind the fact that the organism is a whole and that cell development also responds to changes in the cellular environment of the neighboring cells and of the whole organism.

Only with this preliminary articulation of his program did Harrison realize the productive power of an experimental approach. But only after a few false steps did he discover that merely presenting presumably definitive results did not immediately convince others that they were in fact definitive. At first he set out to prove specific points, an effort that failed to persuade opponents and that must have helped to shape his later ideas about effective methodologies.

One outstanding example of this failure occurred in 1907, when Harrison produced a very direct short paper.[45] Originally read before the Society for Experimental Biology and Medicine, this paper purported to "show beyond question that the nerve fiber develops by the outflowing of protoplasm from the central cells."[46] Harrison had used experimental techniques to transplant into an artificial culture medium a piece of embryonic tissue that normally gives rise to nerves; this constituted the first successful true tissue culture. But the technique, albeit ground-breaking, was not in itself what interested Harrison.[47] Rather, he sought to use this experimental transplantation technique to show how nerve material behaves outside the influence of its normal environment. If nerve fibers formed as usual, then he had demonstrated that they *could* develop without external influences and that hypotheses requiring such external influences (external to the nerve cell and fiber, that is, though they could still be inside the organism as a whole) were not defensible. This rather simple demonstration, Harrison held, provided a supposedly definitive experimental test of the various hypotheses about nerve fiber development: Do nerve fibers form by coagulation of nondifferentiated material, by growth along preestablished bridges, or by protoplasmic outgrowth?[48] Because the alternatives were disallowed, only one possible theory remained, the outgrowth, or neuron, theory. Furthermore, the experimental methods opened the way for addressing other carefully defined questions relating to nerve development, Harrison wrote:

> The possibility becomes apparent of applying the above method to the study of the influences which act upon a growing nerve. While at present it seems certain that the mere outgrowth of the fibers is largely independent of external stimuli, it is, of course, probable that in the body of the embryo there are many influences which guide the moving end and bring about the contact with the proper end structure. The method here employed may be of value in analyzing these factors.[49]

But one absolutely essential element was missing from Harrison's experimental approach, an approach that he hoped to make a definitive one: the use of controls and careful comparison of the experimental with the control specimens. This left him open to charges that his experimental results did not reveal anything about normal conditions. And those charges were indeed presented.[50] Thus others, with different points of view, rejected his conclusions. Direct demonstration and proof by disproof of alternatives did not work because opponents did not admit the results as disproofs.

By 1908, when he presented his summary lecture to the Harvey Society in New York, Harrsion had become an advocate of the experimental approach, with hypothesis testing and experimental techniques, but still without a full sense of the way to use his larger program to make this particular points compelling. Although by then he recognized that experimental results did not always succeed in convincing the opposition, he seems to have felt that what he needed to achieve compelling results was to define his experimental questions more carefully by using isolation experiments and to accumulate more evidence. He asserted that

> although my own work upon the normal development of the salmon and frog has led me to a decided opinion in favor of the cell-outgrowth theory, the attitude of many later investigators showed that we should never be able to obtain evidence from the study of normal development, that would convince everyone alike of the truth of either of the views just stated. A decisive answer to the question, it

seemed to me, could be obtained only by a more exact method of study, i.e., by the elimination, in turn, of each of the two conflicting elements.[51]

Thus he reported the results of two sets of experiments that had eliminated respectively the source of the Schwann cells (after which he found that nerve fibers developed anyway) and the source of the ganglion cells (after which fibers did not develop). This led him to conclude that only ganglion and not Schwann cells are essential for nerve fiber development. But, he continued, "unfortunately for the purpose of devising a clear-cut and crucial experiment, the antithesis between the two views [about nerve fiber development] is not complete. . . . The first experiments of my own . . . were not crucial." Going on to report the results of his tissue culture experiments, he finally summarized that "we have in the foregoing a positive proof of the hypothesis first put forward by Ramon y Cajal," that nerve fibers develop by protoplasmic outgrowth.[52]

Yet at the end, his tone became more cautious with respect to his ability to "prove" his results; instead he discussed placing the neuron theory "upon the firmest possible basis,—that of direct observation." Although he still thought his work had delivered the needed proof, he recognized that others might not think so and that even a seemingly "crucial experimental test" of a hypothesis would perhaps not be definitive. Thus he chose his words carefully, calling the "attractive" alternative "untenable" and concluding, "The embryological basis of the neurone concept thus becomes more firmly established than ever."[53] Firmly established it may have been, but not "proven."

Several factors help to explain Harrison's increasing appreciation of what it takes to convince others who hold different assumptions about appropriate methodologies, problems, or results. First, on the practical level, the experimental test of hypotheses had not worked. Opposition to the neuron theory continued. Given Harrison's commitment to making his work definitive, he had to seek new methods of making his results accepted. Second, Harrison had moved from The Johns Hopkins University Anatomy Department to Yale University, as Bronson Professor of Comparative Anatomy, a joint position in the Sheffield Scientific School, Yale College, and the Graduate School. There he had assumed the position of a biologist rather than a medical school anatomist. This, I suspect, was critical to his changing attitudes. For although he still addressed medical audiences, he also had to communicate with a different general scientific and university audience, which saw the traditional medical problems from a different perspective. This audience regarded the old anatomical problems of delineating how the adult nerves get the way they are from the perspective of such biological problems as establishing the dynamics of how a single egg cell becomes a complex organism, the current rage in biology. This new audience would have been most interested in the general aspects of Harrison's work: in his use of tissue culture, in his ability to isolate parts of an embryo experimentally, in his testing of hypotheses and presenting answers to questions about what happens in development. After his move to Yale, Harrison became more directly concerned with biological processes as well as patterns of development and the anatomical end product. Perhaps most significant, the importance of articulating and making effective a research program had become more urgent to Harrison as the highly acclaimed first University Professor at the traditionally oriented Yale University, since he wished to attract students and to create a research institute of the German style.[54]

Of course, a change of attitude in itself cannot effect results. The fact that his conclusions were still not acceptable to everyone was manifested by the continued publication of opposing views. Therefore, in 1910, Harrison launched a full attack on the opposing nerve development theories that required more than protoplasmic outgrowth from ganglion cells. Specifically, he countered Hans Held's protoplasmic bridge theory. As a result, in 1910 Harrison produced two papers of quite different natures. The first, published in Roux's *Archiv*, provided his final statement within the context and with the style of the descriptive anatomical/embryological debate about nerve development.[55] This paper paralleled earlier experiments to show that nerve fibers grow out into an artificial medium, namely clotted lymph, and that the ganglion cell alone is crucial for nerve fiber growth and development. Furthermore, he cited his results, discussed earlier, that functional activity cannot play a necessary role in nerve development because experimentally inactivated nerves do develop in any case.[56]

The second paper, often cited as Harrison's "crucial" paper, appeared in the American *Journal of Experimental Zoology*, which Harrison edited.[57] This paper drew upon many of the same research results that he had discussed before, but it presented the case for the neuron theory within the context of a *biological* research program. Harrison now outlined his broader research program to uncover experimentally the details and mechanics of embryonic development. He argued within the context of current biological theories about mechanisms and causal features in development, theories influenced especially by Gustav Born, Wilhelm His, Roux, and their contemporaries arguing for experimental embryology, the physiology of development, or "Entwickelungsmechanik."[58] With this paper emerged a fully articulated presentation of the research program that Harrison had been developing. This program soon formed the fruitful basis of an identifiably biological research school centered around Harrison in Yale Unversity's new Osborn Zoological Laboratory.[59]

This second paper of 1910 is an unquestionable masterpiece, one that appears to have silenced even the most stubborn remaining opposition to the neuron theory.[60] By amassing all available evidence from experimental tests, claims by opponents, and counterdemonstrations, Harrison certainly weakened the case for the opposition. Most important, however, he abandoned the claim for a crucial, decisive test between the two major theories—the protoplasmic bridge and the neuron theories. In a sense he abandoned or moved beyond a naive sense of definitive hypothesis testing. Rather, he argued that the two theories were not so obviously opposed as they might seem but were in many ways complementary.[61]

He thus defused opposition and went on to argue for his own research program. He had showed that cellular protoplasmic outgrowth could produce nerve fibers, which he demonstrated were just like normal fibers in the controls; thus he used his controls effectively. "In order to discover the factors which influence the formation of the nerve paths," Harrison wrote, "we must, therefore, in the first instance take into consideration this property of protoplasmic movement. This is of the utmost importance, and any theory of nerve development which fails to do this is sure to be misleading."[62] Further, he claimed that the protoplasmic development into nerve fibers falls "within Roux's definition of self-differentiation, by which is meant, not that the process is entirely independent of external conditions, but simply, as Roux . . . in first defining the concept pointed out, that the changes in the system, or at least the specific nature of the change, are determined by the energy of the system itself."[63]

What these experiments and their interpretation confirmed was the role of protoplasm and cells in nerve growth.[64] Harrison acknowledged that he did not know how the nerve makes its connection with the periphery, but he did indicate how one might explore those connections in future work:

> In studying the secondary factors which influence the laying down of the specific nerve paths of any organism, we are concerned, therefore, primarily with the laws which govern the direction and intensity of protoplasmic movement, and it is the analysis of these phenomena to which students of the ontogenetic and regenerative development of the nervous system must now direct their attention. The present discussion will not have been in vain if it makes clear that the protoplasmic movement concept is no less capable of rational analysis than is development in general.[65]

This was an extremely fertile paper, exciting in it suggestions for future work and in the example it provided in attitude and approach for carrying out that future work. As I have demonstrated elsewhere, Harrison was perceived as a leader in American experimental biology partly because of this work.[66] He had succeeded in making his work convincing—even if not logically definitive.

## Conclusion

I have argued that Harrison's experimental methodology evolved and that it did so in response to both internal and external factors. Inspired by the commitment to making his work conclusive, Harrison's early use of the most basic experimental manipulations gave way to rigorous hypothesis testing using those techniques and ultimately to the formulation of a research program based on an experimental approach. Commitment to experimental techniques, hypothesis testing, and specialization of research program made it possible to achieve results that the emerging larger embryological community could accept.

Acceptance of an experimental ideal—as distinguished from "naturalist" or "observational" methodologies—may well have occurred widely in biology shortly after 1900, as Coleman and Allen have suggested. Harrison's case shows, however, that he at least did not endorse a fully experimental approach in conscious opposition to alternative techniques or approaches. Neither did he adopt experimental techniques or an experimental approach because they were obviously promising and hence a priori tempting. Further, there is absolutely no evidence that he endorsed experimentation at any level because of seeking to emulate the physical sciences or because he rejected the old morphological and descriptive approaches of his predecessors, as Allen has suggested Harrison's generation of biologists did.[67]

What, then, is an alternative explanation for Harrison's adoption of experimentation, and what does this one case study tell us about the emergence of American experimental biology generally? The real problem is in posing the question in this form: it is misdirected, as is the popular question about why early twentieth-century biology became experimental. It makes no sense to ask *why* Harrison, or his contemporaries, adopted experimentation before asking *whether* he adopted experimentation. Asking "Why experimentation?" already assumes that there is a distinguishable differentiation of experimentation from something else, some nonexperimentation. But Harrison's case shows that what we now call experimental biology, or the experimental ideal, or an experimental approach, or whatever, actually emerged slowly. Although Harrison called his work of 1903 and thereafter "experimental," we would have to mean something different by that word to ask why he accepted experimentation at each stage. Experimental science evolved, therefore, and indeed continues to evolve. The type of experimentation Harrison did in 1910 differed from his experimental manipulations of 1903 and after just as the type of experimentation done today in biology laboratories differs from that done at the turn of the century.

Certainly, it is interesting to explore how experimentation changed or to observe that the evolving understanding of what an experimental approach could do seemed to be accelerated shortly after 1900. But that study is interesting because it is symptomatic of a general complex of changes—in scientific methodology and subject matter, in the arts and philosophy, in politics, and generally.

Surely the emergence of special studies such as embryology as separate from medical studies and the emergence and acceptance of various successful specialized *biological* research programs by 1910 must be as fundamental as the incorporation of experimental methodologies by these research programs when experimentation helped them to achieve results. Those biological programs sought to understand both causal and descriptive aspects of life in their own right. The concern with problems considered to be biological and the institutional acceptance of that commitment in the United States—in universities, journals, and research laboratories—involved much more than a wholesale and "almost immediate" endorsement of an isolable experimental ideal in biology.

Around 1900, in fact, there was no such fully developed ideal but only a complex of changing commitments. Even those who thought they were endorsing Roux's experimental program of "Entwickelungsmechanik" did not immediately adopt what they came to endorse as an experimental approach and what Coleman and Allen have termed the experimental ideal. Some of these investigators began with cell-lineage studies which at first simply employed basic experimental manipulations to give the observer control over his specimens. With such noted "experimental biologists" as Wilson, Lillie, and Conklin, for example, we see an evolution, similar to Harrison's, in the style of experimentation thought to be desirable. Morgan's work also demonstrates a parallel evolution from use of experimental techniques in his early regeneration studies to later hypothesis testing and an experimental approach.[68]

Thus, the experimental ideal did not exist as an ideal until investigators realized how experimentation could be used within specialized biological research programs to achieve results. That a number of Americans did produce research programs with experimental methodologies demonstrates their concern with achieving significant results with regard to narrowly defined questions—for which experimentation works well—and the existence of institutional and social support systems for such programs. To produce a recognizable portrait of the emergence and acceptance of American experimental biology that conforms to and seeks to explain the historical data, therefore, it will be useful to draw accurate descriptive pictures and rigorously illustrated analyses of shifting individual and then group assumptions.

NOTES

1. Garland Allen, *Life Science in the Twentieth Century* (New York: John Wiley & Sons, 1975), esp. pp. xix–xxiii.

2. William Coleman, *Biology in the Nineteenth Century* (New York: John Wiley & Sons, 1971).

3. Allen, *Life Science*, p. 34.

4. Allen, *Life Science*, chaps. 2 and 3: "Revolt from Morphology," I and II, esp. pp. 21–72; Coleman, *Biology*, p. 162.

5. This claim is supported by personal conversations with both men.

6. For discussion of ideas of experimentation see: Jane Maienschein, "Problems of Individual Development: Experimental Embryological Morphology," paper delivered at meeting of the History of Science Society, New York, December 1979, and "Crucial Experiments" and "Development" in *Dictionary of the History of Science* (1981). Also Edward Manier, "The Experimental Method in Biology," *Synthese* 20 (1969): 185–205.

7. "Research program" is not used as a theory-laden philosophical term here, but within the context of general usage of the early twentieth century.

8. The published material provides the bases for this study, since I am interested in changes in how Harrison presented his work rather than in the genesis of that work.

9. Ross Harrison, "Die Entwicklung der unpaaren u. paarigen Flossen der Teleostier," *Archiv für Mikroscopische Anatomie* 46 (1895): 500–578; and Moritz Nussbaum, "Zur Mechanik der Eiablage bei Rana fusca," *Archiv für Mikroscopische Anatomie* 46 (1895).

10. For example: Dennis McCullough, "W.K. Brooks's Role in the History of American Biology," *Journal of the History of Biology* 2 (1969): 411–38; Jane Maienschein, "American Biology Comes of Age," paper delivered at The Johns Hopkins University, History of Science Department, January 1981, and Arizona State University, Zoology Colloquium Series, September 1981.

11. Laurence Veysey, *The Emergence of the American University* (Chicago: University of Chicago Press, 1965), esp. pp. 130–31.

12. John S. Nicholas, "Ross Granville Harrison," *National Academy of Sciences Biographical Memoirs* 35 (1961): 132–62, esp. p. 137.

13. Nicholas, "Harrison," p. 132.

14. Ross Harrison, "Moritz Nussbaum," essay (March 16, 1920) in Harrison Collection, Yale University Manuscripts and Archives, New Haven, Conn.

15. Nicholas, "Harrison," p. 134.

16. Edmund B. Wilson went to Naples in 1882–83 and reported the importance of the "new vistas of scientific work" there, as discussed in Thomas H. Morgan, "Edmund Beecher Wilson," *National Academy of Sciences Biographical Memoirs* 21 (1941): 315–42. Morgan spent time there during 1894–95 and said that his Naples experience was a turning point for him scientifically. He became involved in the controversy over half embryos, for example: Thomas H. Morgan, "The Formation of One Embryo from Two Blastulae," *Archiv für Entwickelungsmechanik der Organismen* 2 (1895): 65–71; Harrison, in contrast, spent only a brief time in 1896 and made no reference to the importance of the visit, while he often mentioned Bonn as critical.

17. As demonstrated by the labeling in journals, university departments, and research institutes, for example.

18. Coleman, *Biology*, p. 15.

19. Coleman, personal communication.

20. Interest in evolution was manifested in various ways, not necessarily the same for each of the individuals and not necessarily in Haeckel's extreme sense, which tends to be identified with the period. Careful study of each individual will help illuminate this important factor.

21. For example, Harrison's description of biology in 1914, to be quoted in this essay, implies the acceptability of evolution. In 1922 he wrote that "in my opinion there are no Geologists or Biologists of note who do not believe in the correctness of the evolution theory" (Harrison to Mabel Moore, December 8, 1922, Harrison Collection). And numerous exchanges with his evolution-minded friend Edwin G. Conklin reveal Harrison's belief in evolution. Yet Harrison never discussed evolution in his research papers. Neither phylogenetic nor other evolutionary factors had a place in laboratory work, according to Harrison, because it was impossible to study evolution "scientifically." This interpretation of Harrison's attitudes was reinforced by personal conversation with Evelyn Hutchinson, February 2, 1977.

22. See Jeffrey Werdinger, "Embryology at Woods Hole: The Emergence of a New

American Biology," Ph.D. dissertation, Indiana University, Bloomington, 1980. Werdinger discusses what he terms the "internal" and the "external" critiques of previous phylogenetic traditions in embryology. I do not mean to suggest that medicine is antievolutionary in any way; rather evolution provided no essential constraints or directive influence for medicine around 1900.

23. For example, in 1914 Harrison was offered a position as head of the anatomy department in the medical school at Johns Hopkins; in 1924 he was offered a similar position in the newly organized Columbia University Medical School. Instead he chose to stay at Yale as Sterling Professor of Biology. Nicholas, "Harrison," esp. p. 146, discusses other medical-biological connections during his career at Yale.

24. Ross Harrison, "The Osborn Memorial Laboratories, I: The Zoölogical Laboratory," *Yale Alumni Weekly* 23 (1914): 668–69.

25. Nicholas, "Harrison," p. 146.

26. Harrison, "Die Entwicklung." While not concerned with tracing the course of individual cells, Harrison concentrated on cellular changes in later embryonic stages rather than on the germ layers Haeckel and others had emphasized. In this he was like his contemporary American cell-lineage investigators.

27. Harrison, "Die Entwicklung," p. 527. Original: "Über die Entwicklung der Nerven in den unpaaren Flossen, vermag ich nichts Wesentlicher anzugeben. Mit Hülfe der gewöhlicher embryologischen Methoden sind die Einzelheiten der Entwicklung nicht ausfinden zu machen."

28. I feel certain that Harrison would have known of this work because of his study in Germany and his time spent in Woods Hole and at the Marine Biological Laboratory, Woods Hole, Massachusetts. He probably knew of Roux's ideas, therefore, but found other work more interesting, with different goals and different assumptions. Undoubtedly he knew of Roux's introduction to his journal, *Archiv für Entwickelungsmechanik*, translated by William Morton Wheeler as "The Problems, Methods, and Scope of Developmental Mechanics," *Biological Lectures* (Boston: Ginn & Co., 1894), pp. 149–90.

29. Gustav Born, "Über Verwachsungsversuche mit Amphibienlarven," *Archiv für Entwickelungsmechanik* 4 (1896): 349–465; 517–623.

30. Ross Harrison, "The Growth and Regeneration of the Tail of the Frog Larva," *Archiv für Entwickelungsmechanik* 7 (1898): 430–85, esp. p. 430.

31. Harrison, "Growth and Regeneration," p. 430.

32. Harrison, "Growth and Regeneration," p. 435.

33. Ross Harrison, "Experimentelle Untersuchungen über die Entwicklung der Sinnesorgane der Seitenlinie bei den Amphibien," *Archiv für Mikroscopische Anatomie* 63 (1903): 35–149.

34. Harrison, "Experimentelle Untersuchungen," pp. 142–43.

35. The Harrison Collection at the Yale University Manuscripts and Archives has only a few items such as notebooks from Harrison's earlier years, but even the material available reflects nothing different than the publications. It would be desirable to know more about why this change came in 1903. Maybe Harrison had simply become aware of other work and theories, maybe the impending 1904 anatomy meeting at Jena provided the crucial stimulus; the reasons simply are not clear.

36. Sally Wilens, Introduction to *Organization and Development of the Embryo*, by Ross Harrison (New Haven: Yale University Press, 1969), p. xiii.

37. Sheath cells could be seen, using microscopic techniques, surrounding the nerve axon as soon as anatomists could see nerves. They could not tell, therefore, whether the sheath cell caused (that is, was antecedent to), was caused by, or was completely independent of the fiber it surrounded.

38. Ross Harrison, "Nachtrag zu der Diskussion zu den Vorträgen von Schultze and v. Kölliker," *Verhandlung d. Anatomische Gesellschaft, 80 Versammlung*, Jena, 1904, p. 52. It would be enlightening to know more about this meeting. The papers were fairly descriptive, presenting data to support the preferred conclusions. But reports suggest that theory and methods were also vigorously discussed.

39. There are, of course, other meanings, some philosophically loaded, but this seems to have been what Harrison meant.

40. Ross Harrison, "An Experimental Study of the Relation of the Nervous System to

the Developing Musculature in the Embryo of the Frog," *American Journal of Anatomy* 3 (1904): 197–220, esp. pp. 199–200.

41. There was considerable discussion in the 1890s, especially in German anatomical journals, about the distortions caused by manipulations. Oscar Schultze and Hans Held were leading disputants on the issue of nerve fiber development in particular.

42. Ross Harrison, "Further Experiments on the Development of Peripheral Nerves," *American Journal of Anatomy* 5 (1906): 121.

43. Harrison, "Experimental Study," p. 216.

44. Harrison, "Experimental Study," p. 218.

45. Ross Harrison, "Observations on the Living Developing Nerve Fiber," *Anatomical Record* 1 (1907): 116–18.

46. The Society for Experimental Biology and Medicine, according to discussions in the Harrison Collection at Yale University, was an attempt to bridge those two subjects by coordinating experimental studies in both.

47. Jane Oppenheimer, "Ross Harrison's Contributions to Experimental Embryology," *Bulletin of the History of Medicine* 40 (November–December 1966): esp. p. 525 establishes the greater importance of Harrison's questions than his techniques.

48. For further discussion see Jane Maienschein, "Ross Harrison's Crucial Experiment as a Foundation of Modern American Experimental Embryology," Ph.D. dissertation, Indiana University, Bloomington, 1978, esp. chaps. 1 and 3.

49. Harrison, "Observations," p. 118.

50. For example, Hans Held, *Die Entwicklung des Nervengewebes bei den Wirbeltieren* (Leipzig: Verlag Johann Ambrosius Barth, 1909).

51. Ross Harrison, "Embryonic Transplantation and Development of the Nervous System," *Anatomical Record* 2 (1908): 391.

52. Harrison, "Embryonic Transplantation," pp. 397 and 409.

53. Harrison, "Embryonic Transplantation," p. 410.

54. This interest is reflected in various archival sources in the Harrison Collection at Yale University and is summarized in an undated document, "Program for an Institute for Research in Experimental Embryology and Related Fields."

55. Ross Harrison, "The Development of Peripheral Nerve Fibers in Altered Surroundings," *Archiv für Entwickelungsmechanik* 30 (1910): 15–33.

56. Harrison, "Development," p. 30.

57. Ross Harrison, "The Outgrowth of the Nerve Fiber as a Mode of Protoplasmic Movement," *Journal of Experimental Zoology* 9 (1910): 787–846.

58. Frederick Churchill, *Wilhelm Roux and a Program for Embryology*, Ph.D. dissertation, Harvard University, 1966, esp. chap. 2.

59. Maienschein, "Ross Harrison's Crucial Experiment," chap. 4.

60. Even the archopponent of the neuron theory, Held, published no further rebuttals on the subject.

61. Harrison, "Outgrowth," p. 830.

62. Harrison, "Outgrowth," p. 832.

63. Harrison, "Outgrowth," p. 833.

64. Harrison, "Outgrowth," p. 840.

65. Harrison, "Outgrowth," pp. 840–41.

66. Maienschein, "Ross Harrison's Crucial Experiment."

67. Allen, *Life Science*, esp. pp. xvii–xix.

68. Jane Maienschein, "Shifting Assumptions in American Biology: Embryology, 1890–1910," *Journal of the History of Biology* 14 (Spring 1981): 89–113.

# Signs of Dominance: From a Physiology to a Cybernetics of Primate Society, C. R. Carpenter, 1930–1970

*Donna Haraway**

> A *picture* held us captive. And we could not get outside it, for it lay in our language and language seemed to repeat it to us inexorably.
>
> —Ludwig Wittgenstein

> If an ape peeks in, no apostle peers out.
>
> —Georg Christoph Lichtenberg

## Introduction

At the height of the production of Salk polio vaccine in the 1950s, the United States imported over 150,000 rhesus monkeys per year for industrial and research purposes. In addition to using rhesus material in the manufacture of many biologicals, six pharmaceutical companies needed fresh kidney tissue from over 120,000 monkeys to produce

*I am indebted to the gracious and knowledgeable librarians in the special collections of the Yale University School of Medicine, the National Academy of Sciences, and the Pennsylvania State University. Sherwood Washburn opened his personal correspondence files and generously participated in interviews. Several colleagues read this paper and contributed excellent critiques. Thanks are due especially to Constance Clark, Stephen Cross, Nancy Hartsock, Rusten Hogness, Robert Kargon, Camille Limoges, Jaye Miller, and Philip Pauly. I am grateful to Roberta Yerkes Blanshard for permission to quote from her father's papers. Students in seminars at The Johns Hopkins University History of Science Department and L'Institut d'histoire et de sociopolitique des sciences of the Université de Montréal commented extensively on drafts and lecture versions of theses in this paper. Board of the History of Consciousness, University of California at Santa Cruz.

vaccine; and about one thousand laboratories used 25,000–40,000 of the animals for a wide range of biomedical and behavioral research. In 1955 the Indian government, in response to internal religious and political complaint about inhumane conditions of trapping, shipment, and use of rhesus monkeys, declared an embargo on their export. The events set in motion in the U.S. scientific community by that embargo were part of a broad reawakening of interest in the role of nonhuman primates in the management and interpretation of human life.[1] This essay concentrates on the period from 1930 to 1970 and on the work of Clarence Ray Carpenter (1905–75) in order to tell the story of the production of a primate science as part of the theorizing of bodies, societies, and populations as problems in the rationalization and maintenance of dominance. The rich ambiguity of the scientific principle of dominance and the political practice of domination is also at the heart of this essay.

Carpenter was an important figure in the development and extension of the theory of the dominance hierarchy as the organizing axis of organic social structure. His work is essential to understanding the extension of laboratory-based comparative psychological and physiological work on the dominance principle in the 1920s and to the study of primate societies in the field. In addition, Carpenter's writings illuminate the relation of the theory of organic dominance hierarchy to the definition and control of human problems—especially the enforcement of "cooperation" and avoidance of "conflict" in production and reproduction—within a scientific orthodoxy that insisted on a proper distinction between nature and culture and between natural and social sciences.[2] Carpenter's work took shape within a set of rules forbidding the Spencerian form of naturalization of human history.[3] The animal dominance hierarchy could not be directly applied to human life, nor could any other specific mechanism in animal groups give direct insight into the academically well-defended realm of "culture," after the late 1930s. Extrapolation was redundant. The logic of domination that this paper explores was common to both the natural and the social sciences and was rooted in the multiple concrete processes that constructed the objects of analysis themselves, whether nature or culture, animal or human being, as embodiments of particular, scientifically validated principles of order. It is the task of this essay to dissect these rules of order as they have appeared in some primate bodies since 1930.

### *Dominance*

The concept of dominance and a guiding logic of domination are made explicit by participants in the practice of primatology at three levels relevant to the theme of this paper: (1) in the development of practices and principles of conservation as resource management and the more focused control of animal supply providing a primary material foundation for the work; (2) in the engineering of a system of laboratories and international fieldwork, the structure of which clarifies the social relations sustaining the production of the science; and (3) in the elaboration of a knowledge which repeatedly and insistently constructs nature as a problem in systems of control. The first two levels form a briefly developed backdrop for the third, which is the center of this paper. But it would be a mistake to consider the questions of laboratories and supply of animals and the social relations and ideologies of a larger society as the "social" levels of analysis, while the third level, knowledge, either properly proceeds by technical and theoretical refinement or unfortunately subordinates "technical" progress to ideological servitude.

The technical and social dimensions of the constitution of primate science are not merely intertwined; they are identical.[4] Techniques themselves are intimately social products; this is not a contamination of science, but a condition of its production. Primatology does not escape this condition. This paper will not be about bad or ideological science, but about one window into the constitution of scientific knowledge and objects of that knowledge, in this case societies of monkeys and apes. To construct objects of knowledge by the practice of a logic of domination does not produce false knowledge, but particular knowledge. It might be possible to know something else instead. The order of Primates might yet be part of a radical deconstruction and reconstruction of the rationally known body. But this paper restricts itself to attempting to show how, where, and when an area of primatology was built according to a logic of domination operating on the three levels named above.

First, in the context of the embargo of 1955, dominance meant the assertion of U.S. control over the supply "of one of our major tools of research, namely, the living animal."[5] Primates other than human beings do not occur naturally in North America; the sources of supply for Old World primates are Asia and Africa. Restrictions, other than those self-imposed by the scientific community, on the primate trade have occurred repeatedly in the twentieth century. Even more ominous for U.S. assurance of an uninterrupted supply of a basic raw material in the production of science, primates have been increasingly threatened with extinction in their natural habitats. Regulation of trade and international coordination of conservation were essential to control of supply and form a central part of the larger story of primate studies.

But as the chairman of the Division of Biology and Agriculture of the National Academy of Sciences–National Research Council, Paul Weiss, observed at the conference called to deal with the 1955 crisis in the monkey supply, "It occurred to me right from the start that conservation, that is, merely the conservative aspect is not really the only answer, but what we have is a balance sheet between production and consumption ...

You can raise the balance sheet by increasing production . . . Irrespective of how large the global supply may be, what counts here is merely what we can get our hands on."[6] Weiss stressed that, fundamentally, foreign sources of supply "are not entirely under our control. They may be under the control of foreign relations, and they may dry up without our being able to do anything about it on the spur of the moment. We may be facing a national emergency one day which is entirely beyond our control at least on the short range."[7] His reiteration of the word *control* and polite evocation of the international context of the acquisition of raw materials for science and industry by developed countries after World War II highlight a basic kernel of domination in the practice of primate science.

Weiss pointed out the consequences of his argument—the desirability of expanded U.S. production in this hemisphere of primates for industry and science. In that context, the second level of dominance important to the theme of this paper emerges: construction of a system for producing nonhuman primates fit to meet the growing demands. A principal demand was that the primate laboratory serve as a pilot plant for human engineering. That goal was rooted in the earliest plans and achievements of primatology in the United States. Robert M. Yerkes (1876–1956) spent a lifetime of scientific service building a research program and institution centered around the chimpanzee as a model for human life. Yerkes's holistic psychobiology extended from physiology to behavior to primate sociology. His purpose was to build a comprehensive experimental primate science, modeled on experimental physiology, which could serve as the foundation for rational social management, as physiology provided the rational ground for medical therapeutics. Yerkes expressed the spiritual foundation of the primate laboratory frequently and forcefully:

> It has always been a feature of our plan for the use of the chimpanzee as an experimental animal to shape it intelligently to specification instead of trying to preserve its natural characteristics. We have believed it important to convert the animal into as nearly ideal a subject for biological research as is practicable. And with this intent has been associated the hope that eventual success might serve as an effective demonstration of the possibility of re-creating man himself in the image of a generally acceptable ideal.[8]

In 1929 Yerkes finally obtained the material foundation for his dream in a 10-year $500,000 grant from the Rockefeller Foundation for a comprehensive primate laboratory, Yale University's Laboratories of Comparative Psychobiology. The Yale laboratories had a three-part structure, with specialized physiological and psychobiological facilities in the medical school, in New Haven; a breeding station with about 40 animals in Orange Park, Florida; and sponsorship of field studies of wild primates wherever opportunity arose. Under Yerkes's sponsorship, Carpenter conducted the field investigation of howler monkeys in the

Panama Canal Zone that led to concepts and field techniques that this work will explore in detail.[9]

In addition, Carpenter's connections with the network of powerful scientific men, which Yerkes helped to build and maintain, sustained the younger man's own effort in the late 1930s to establish a semi-free-ranging colony of rhesus monkeys and gibbons on Cayo Santiago, off Puerto Rico. Carpenter had that opportunity partly as a result of a limited embargo imposed by the Indian authorities on the export of rhesus monkeys in 1938, when the U.S. demand was for only about 1,000 animals per month. The Cayo Santiago monkey production experiment was allowed to deteriorate in the 1940s and 1950s, but a ragged colony still existed in 1955 and will enter into our story often. Weiss, again at the meeting under the auspices of the National Research Council in the face of the 1955 embargo, said:

> We have had a pilot plant experiment going in Puerto Rico, and I am delighted to have some of the representatives of this very successful experiment here with us today [in particular, Carpenter]. . . . An island has been stocked with monkeys. If it can be done once, it can be done again. If it can be done again, we might set our sights higher than merely producing monkeys.[10]

The sights were on a comprehensive scientific achievement relating conservation and supply, production, and theoretical and methodological development; in short, a "mature" primatology to take its place in the complex household of biological and social sciences, which defined and helped enforce the human place in nature and culture in postwar America. The 1955 conference was a first step, culminating in the seven National Institutes of Health Regional Primate Research Centers. The first new center opened in 1960 in Oregon, and the seventh began operation in 1965. By the early 1970s, the regional centers, each associated with a major university, housed 8,000 primates belonging to 45 species, accommodated 400 scientific investigators and over 700 supporting staff, spent about $15 million per year, and directly produced around 550 scientific papers and 20 books annually. Established by a Public Health Service act to pursue basic and applied research for solution of human health and social problems, these centers represented the fulfillment of Yerkes's dream and concrete promotional work he had begun before World War I.[11] His own Orange Park chimpanzee laboratory was transferred to Atlanta, Georgia, and became associated with Emory University. In cooperation with this institution, the Yerkes Regional Primate Research Center, after retirement from Pennsylvania State University in 1968, Carpenter worked to initiate the first Ph.D. program in primatology in the U.S. at the University of Georgia.

The story of primate research laboratories as pilot plants for human engineering—from Yerkes's physiological-psychobiological center at Yale, to the Cayo Santiago breeding colony, to the national system of specialized facilities, each with a differentiated mission in primate research relating to reproductive biology, population control, mother-infant relations, mental retardation, social organization, communication systems, infectious and degenerative diseases, psychiatric disorders, learning, and neural biology—illustrates a key transformation in biological research in the twentieth century. This transformation was central to the development of the relations of dominance in scientific production, on the levels of scientific social organization and of conceptual-technical knowledge through which the natural world is constructively known. Again Weiss recognized the issue: "It became clear to me along with some of my friends that science and biology, particularly in this country, is (sic) entering a new phase, a phase which I might indicate is a transition from the workshop or the individual worker to something which involves the mass application of an industry."[12] The early work of Carpenter, alone in the field in Panama, and the first efforts of Yerkes to establish primate studies institutionally represented a workshop stage of biology. The automated and specialized laboratories of the Regional Primate Research Centers marked a full industrial development of the early pilot plants, with very different production processes directed toward human engineering. By this stage, Carpenter and Yerkes were ritual citations in primate publications, but no single individuals could play their former roles. By 1970, Carpenter himself was a minor figure struggling to advance his own enterprise in severe organizational and theoretical competition with other productive centers of primatology.[13]

*Dominance and Functionalism*

The intimate connections of knowledge and the system which produces it lead to a third level of dominance in the primate story: scientific language, theory, and technique. Primatology has built a science around the problems of integration, coordination, and control. The science of primates is concerned with both whole individual organisms and whole social groups, as well as with populations, species, and other objects of evolutionary and ecological biology. As a branch of life science, primate studies are about function, structure, and their dynamic interrelation; i.e., primatology is a functionalist discourse, concerned with the organization of wholes, their transformations over time and space, and their mechanisms of coordinated action. The teleology built into functionalism in life and human sciences is one of control; i.e., the establishment and maintenance of stable order and rule-governed transition. Functionalism is interested in action rooted in structure; functionalism is about governed motion of bounded wholes. Functionalism studies logics maintaining dynamic and complex structures. From about 1930 to 1970 there was a major transformation in the type of functionalism informing biological practice and theory. This transformation has profoundly altered the distribution and interconnections of physical, life, and human sciences. This essay will argue that the organismic, physiological functionalism that unified biology in the 1930s has been reconstituted since World War II as a cybernetic functionalism, ordered by communications, information, and systems theories on both practical technical and theoretical levels. A corollary of the transformation from physiological to cybernetic functionalism is that the logic of control of the object known has changed from organic medical therapeutics to systems engineering. These control logics refer both to the internal laws of a functioning whole independent of obvious human purposes and to rules governing effective human intervention in and appropriation of the object of study. The structure of control logics makes no distinction between basic and applied science.

Further, a control logic can be part of a system of domination based on practices of coordination and management, i.e., the scientific way, and not on practices of direct force and traditional authority. Both physiological and cybernetic functionalism have been tools of *rational* domination, precisely the root of their power ideologically and technically: they work; they function. Functionalism is a logic for the mediation of domination through self-sustaining processes, not a logic of direct visible command. For example, even the animal dominance hierarchy was constructed as a mechanism of finely modulated integration rarely operating by simple force. Finally, the transformation of the productive system of biology from workshop or small manufacture to highly automated mass industry corresponds strikingly to the conceptual-technical reformulation of functionalism and its associated control logic from the physiological animal-machine to the cybernetic system.

This thesis could be argued from many vantage points in biology, perhaps most readily in telling the history of genetics and cell biology or of neural biology.[14] But the focus on primatology has the double advantage of highlighting the question of the human being's place in nature and the complex relations of natural and social sciences. Every science must constitute its object of study; indeed, that is its most fundamental achievement. Primates have been constituted as very curious objects; here the problem of knowing what might be distinctively "human" or reliably "natural" has been acute. Primates have been engineered as tools in the constitution of human and nonhuman nature. Nature is fiction and fact; i.e., made in practice, a cultural production. From the time of Linnaeus, the effort to know the primates has been an effort to construct a natural order.[15]

*Carpenter's Tools*

C. R. Carpenter's writing, promotional activities, funding pattern, and institutional bases are coordinated guides to the reorientation of primatology from a physiological to a cybernetic functionalism. In both forms, his theory and practice of primatology according to a logic of dominance show forcefully the scientific operation of domination as therapeutics, management, and engineering, in contrast to brute hierarchy and traditional authority. Carpenter began his scientific life during the height of physiological functionalism in the late 1920s and early 1930s, in the Stanford University laboratory of Calvin P. Stone. Continuing work begun in his master's thesis under William McDougall, at Duke, Carpenter obtained his doctorate at Stanford in psychology (1932) for studying the effect of gonadectomy on the sexual activity of male pigeons in mated pairs.[16] The doctoral work was funded from Stone's grant from the Committee for Research in Problems of Sex, a National Research Council, Rockefeller Foundation–supported granting body chaired by Yerkes from 1922 to 1947.[17] The interrelations among reproductive and neural physiology and psychobiology, or behavior, were principal concerns of Carpenter's mentors. Formed in physiology and comparative psychology, from the beginning Carpenter viewed sexual activity as the unifying locus of individual organism and organic society. He began from the widely held premise that societies of higher animals could only be explained in terms of the bionomics of sex: the basic forces of social order—cooperation and competition—must at root be aspects of sexual interaction.[18] Together sex and mind were not only the principal objects of study for endocrinologists, neurophysiologists, and psychologists, but they were believed to constitute the material foundation of organic social integration, as well as the greatest threat of disintegration. Sex and mind were thus the keys to scientific control of life, the fundamental object produced by biology.

Investigations of the form of motor activity patterns, of the intensity of motivation toward satisfaction of sexual drives, of the role of glands of internal secretion in social behavior, and of the specific nature of incentives sufficient to evoke particular behavior patterns: all these made up the study of coordination and integration of the parts of the individual organism and the minimal social group, the mated pair, within the framework of physiological functionalism. Behavior was like morphological structure; both were characterized by form and function and subject to pathology. The defect experiment, like that of Carpenter's gonadectomized pigeons, was the appropriate scientific manipulation to reveal the physiology of individual and of society. In studying sex, the procedure was to cut out glands and organs; in studying mind, scientific procedure dictated altering or removing the head. This logic recurred repeatedly in Carpenter's work. The biomedical context of this research was critical to the explicit connections between the physiology of integration and the sociology of cooperation. Sexual behavior was a privileged handle to the theoretical understanding and therapeutics of natural cooperation ordered by male-female dominance and male-male competition. The physiological structure of coordination was produced by mechanisms of domination and competition among organic parts. The physiological animal-machine generated harmony from the materials of conflict. It was all a question of the laws of balance of opposing forces to form a whole.

Carpenter maintained and deepened his physiological point of view in his first work on primates. Field study of wild howler monkeys in the Panama Canal Zone, on Barro Colorado Island, was made possible by a National Research Fellowship. As a postdoctoral student attached to Yerkes's Yale Laboratories of Comparative Psychobiology, Carpenter spent a total of 12 months in the field from 1931–34, 8 months of that time watching howlers. The howler monkey field journals and published monograph merit detailed examination (sec. 2) in the context of the importance of organismic theories in the 1930s in the U.S. Section 2 will also develop the theme of the determining influence or ornithology, developmental physiology, and community theory ecology on the foundation of primatology as the field study of wild populations.

Carpenter's next major research, conducted from a base as lecturer at Bard College of Columbia University, was the study of gibbons in Thailand as part of the Asiatic Primate Expedition of 1937. At this point in his work, he introduced concepts and practices critical from the later perspective of a transformation from physiology to cybernetics. In particular, Carpenter adopted a philosophical and linguistic theory of signs; i.e., *semiotics* as practiced at the University of Chicago, and sociological-psychological field theories of complex small group structures, especially *sociometry,* to explain the pattern and boundaries of primate social organization. Far from primate studies serving as a reservoir of resources for the practice of biological reduction of human sciences, primatology has been pervasively determined by borrowings from human social science. That fact explains some of the ease with which strategies of biological reductionism can be found when desired; the biological disciplines are already built like other contemporary functionalist discourses. Carpenter's gibbon work will receive detailed consideration in section 3. The same section will explore the establishment of the rhesus colony on Cayo Santiago and the experimental manipulations of social groups that confirmed Carpenter's use of semiotic and sociometric analyses, but still within a physiological-functionalist framework.

World War II marked a critical turn in biological work, including primatology. Carpenter's part in making training films for tropical warfare, directing the Research Section of the Biarritz American University in France, and in planning postwar German youth reeducation programs all reinforced preexisting interests in educational and communications technology. Quickly, during and after the war, on practical and theoretical levels, Carpenter transformed his approach to questions of integration and coordination to accommodate developments in cybernetic systems theories. These reorientations became evident in the changed conditions of primate research from the 1950s. Section 4 will probe the cybernetic structure of that area of primatology represented by Carpenter and his close associates.[19] From 1940 to 1968 Carpenter was on the faculty of the Psychology Department of Pennsylvania State University. After World War II, half his time was committed to the Instructional Film Research Program at Pennsylvania State and to related activities in local, national, and international development of technologically sophisticated communications systems to achieve social "integration."[20] In this later period, the connection of Carpenter's primate work with his other activities was important in the production of knowledge and technology of control sytems. From 1968 until his death in 1975, Carpenter was at the University of Georgia in the double capacity of consultant to the president and research professor in psychology and anthropology.

## 2. The Physiology of Dominance, 1929–1937

> Physiological investigations of recent years show that in all except the simplest organisms . . . physiological integration results from the establishment of a physiological dominance, a leadership or government in the protoplasm composing the organism.
>
> —Charles Manning Child, 1928

In the 1930s Carpenter's work focused on pigeons and chimpanzees in psychological laboratories attached to the programs in reproductive physiology funded by the Committee for Research in Problems of Sex; howler and spider monkeys at field stations created near the Panama Canal and on land owned by the United Fruit Company; gibbons and orangutans in Thailand and Sumatra before strict conservation regulated scientific hunting (collecting); gorillas in the San Diego Zoo; rhesus monkeys on a provisioned island attached to a colonial medical school affiliated with Columbia University; college men in dormitories; and U.S. married couples in questionnaires. In all these cases, Carpenter sought to understand governance, the physiological structure of control, through studying sex as steersman. The result was a bionomics of organic social groups that was part of the development of population biology as a key discipline in the management of human life according to medical logic. This section will follow Carpenter from his postdoctoral work to the beginning of a last major primate expedition before World War II, the ship-borne Asiatic Primate Expedition of 1937. The purpose of this section is to show the foundation of the laws of dominance in the logic of physiological functionalism. This logic underlay the biologies and technologies of precybernetic self-regulating systems. After World War II, governance would have more to do with the communications systems that made intercontinental air traffic possible than with the steering of oceangoing ships. The primate story curiously follows these transportation questions.[21]

### *The Beginnings of a Sexual Bionomics of Primate Society*

With the end of his doctoral thesis work in sight, Carpenter sought the help of his Stanford advisers, Stone and Lewis Terman, in obtaining a National Research Fellowship to work in Robert Yerkes's laboratory "investigating some aspect of sex behavior in primates."[22] First under McDougall at Duke and then under Stone, Carpenter studied the physiological basis of the "monogamic relation" of pigeons and concluded that comparative psychology needed greater attention to social responses of animals. He believed that "the animal has been considered primarily as a reacting organism to a stimulus situation. Hence, many workers have lost sight of the fact that the animal itself is a potent stimulating situation for the other members of the group."[23] Carpenter grounded his study of social responses in sexual behavior; here the concerns of reproductive physiology and comparative psychology came together on questions of the utmost social importance for human beings. In sexual bonds, both the origin and cohesion of social bonds might be found. Yerkes was far from indifferent to the application from the promising Stanford Ph.D. In Carpenter he believed he had found the man who could help realize his program of establishing an experimental sociology of primates,[24] with its heart in scientific study of family relations as a basis for general social therapeutics.

The sketch Carpenter sent the National Fellowship Board of his proposed postdoctoral work reaffirmed that hope; Carpenter included in three possible research problems the essential elements of the unified vision of comparative psychology and reproductive physiology focused on the foundation of social life in animals. Here was the place to found a science of organic coordination on the social level. In his first prospectus Carpenter proposed to study social reactions in animals by "not[ing] especially the attachment of one animal for another of the opposite sex." He would study posture, facial expression, and motor action patterns of chimpanzees in response to each other in an effort to measure the strength of social, especially sexual, attachment.[25] The traditional method of comparative psychology, the multiple choice problem, was ready to hand for measuring social bonds. Extending the concern with social cooperation in the second prospectus, Carpenter proposed to set up a movable chain with a food reward, which required the coordinated effort of more than one champanzee to procure the goal. The third problem proposed to extend to monkeys work that had been carried out on small laboratory mammals, poultry, and pigeons, i.e., correlation of sexual behavior and physiological state before and after castration. The design was similar to the one Carpenter followed on his doctoral pigeons. Behavioral characteristics, metabolic rate, and tissue fluids were all to be carefully measured. Carpenter, as well as men like Stone and Yerkes, made no distinction between behavior and other physiological movements in a dynamic organism. The "action patterns" of behavior must be as carefully scrutinized as the secretions of glands for their characteristic form and function. Sex neatly revealed the whole physiology—from body, to behaving organism, to social group. And sexual physiology was that area where the relations of dominance and subordination, activity and passivity, initiative and receptivity, and stimulus and response were most evident and crucial.

Carpenter expected to do his research in the usual laboratory environment of psychobiology and physiology, but he was wrong. Vigilant for opportunities to foster naturalistic field investigations to complement laboratory research, Yerkes had been looking since 1929 for someone to follow up a rich chance.[26] On February 1, 1939 his friend Frank M. Chapman, the founder of the Bird Department at the American Museum of Natural History, who worked extensively on Central American birds from a base on Barro Colorado Island, in the Panama Canal Zone, wrote Yerkes about the wonderful opportunity to study howler monkeys on Barro Colorado. "In the trees nearly over my house there are at the moment seventeen howlers, three of which are carrying young—a unique opportunity here to study the individual, the family, the clan, the interclan relations under a natural but controlled environment."[27] In his capacity as a first-rate field ornithologist Chapman was crucial to Carpenter's initial work and so to the development of field methodology and fundamental concepts in primatology. The study of birds was second only to laboratory-based physiology and comparative psychology in determining the inheritance of primate studies.[28] Since Carpenter's importance lay precisely in the reworking of this inheritance in producing a field biology of nonhuman primates, it is worth following the bird connection further, for there is where lab and field come together.

## *From Birds to Monkeys*

The laboratory pigeon has already entered the story as the vehicle for examining the components of sexual drive. Carpenter expected to extend that work to chimpanzees.[29] He knew the earlier work on pigeons and domestic fowl, which was full of suggestions for a psychobiology of social groups extended from laboratory to field. At the end of the nineteenth century, Charles Otis Whitman, first director of the Marine Biological Laboratory of Woods Hole and chairman of the Department of Zoology of the University of Chicago, and early in the twentieth century, his student Wallace Craig studied pigeons in ways that lessened the distance usually maintained to exist between American comparative psychology and Continental ethology before the 1950s. Whitman, cited favorably by Konrad Lorenz in the 1930s, carefully studied motor action patterns of pigeons as morphologically differentiated forms useful in making taxonomic judgments. Neither Whitman nor Craig was guilty of the usual charge that comparative psychologists neglected varietal and specific differences in behavior pattern in the homogenizing drive to study learning in the laboratory. Both were committed to the evolutionary study of instinct as a problem in functional morphology. Further, Craig studied calls of pigeons as means of social control in groups; his study was important to Carpenter 25 years later.[30]

Besides the study of motor patterns as a problem in functional evolutionary morphology and of vocalization as a mechanism of internal social control, work originally focused on birds affected primate studies in three interlocked and basic areas in the 1930s: territoriality, population biology, and the animal sociology of cooperation and competition. These areas in turn were part of a general physiological functional approach to organismic wholes; Carpenter incorporated all of them into his howler monkey study.[31] In the theory of territoriality, the basic point was the description of territory in behavioral terms; the space of territory was both physical and psychological. In one of its postulated functions, territory was the space of successful reproduction. Behavior, not physical boundaries, marked off territory. Behavior that differentiated space—such as bird-song, scent-marking, aggressive display, or periodic travel to particular places—was a fundamental mechanism of integration of animal societies. Mapping space by mapping behavior became an operation appropriate to social physiology.[32]

### *Counting*

Population biology has many roots, but those that concern us here are the common runners between bird and primate studies. Chapman reenters the story in his encouragement of Carpenter in a simple operation: counting howler monkeys carefully, making a complete census of the howlers of Barro Colorado Island. This operation was essential in developing a bionomics of organic primate society. Since 1905 American economic ornithologists had conducted breeding bird censuses, and in the middle 1930s the National Audubon Society inaugurated a project of quantitative breeding bird censuses. These procedures produced much more precise awareness of territory size fluctuations, fluctuations of population density from year to year, and the influence of slight changes of vegetation.[33] Ornithologists like Chapman were keenly sensitive to the importance of a census as a foundation for basic biological statements. Two kinds of such statements concern us here: population forecasting and descriptions of the growth form of a population. In the period before World War II, both kinds of statements were part of the natural economy of a particular sort of physiological body, i.e., the interbreeding aggregations of animals of the same species, which also sometimes formed groups called societies. It was a natural economy with a definite purpose: population management conducted according to a therapeutic logic of health. A census was not just a quantitative description, but was related to the problem of prediction, to the discovery of the laws of population growth, decline, or stability in particular conditions. The population was considered to be a biological body that could be in a state of health or pathology like any other organism with a history of organic development, maturity, and possibly death. The field censuses or organisms before World War II were another aspect of physiological functionalism.[34]

*Cooperating*

Like territoriality and population biology, animal sociology of cooperation and competition was not limited to studies of birds in the 1920s and 1930s. But the bird work was a conspicuous thread in the fabric that Carpenter further embroidered. Two names are inescapable: Thorlief Schjelderup-Ebbe, the Norwegian who worked in France, and W. C. Allee, at the University of Chicago. Here we encounter the central concept of dominance and its role in physiological functionalism. Schjelderup-Ebbe was credited with the discovery that birds were organized into social hierarchies by a strict dominance chain, or pecking order. Studying over 50 species of birds, he thought he had determined that "despotism is one of the major biological principles."[35] The question of the dominance hierarchy as a mechanism of social coordination became a major preoccupation of zoologists and comparative psychologists in America in the late 1930s. A colleague of Yerkes and important comparative psychologist, Meredith Crawford, could correctly write in 1939 after a decade of experimentation on the dominance hierarchy in nearly every conceivable animal that could be induced to move:

Exploration of the significance of the concept of dominance has hardly begun, since little is known about how dominance may influence all sorts of social interaction. Also, the factors which determine dominance status are yet to be fully elucidated. Among the primates the subject is particularly fascinating, since there the relation between dominance and sex behavior is perhaps most striking. Full exploration of this relationship may suggest solutions of many baffling problems of sex attitudes and perversions as well as indirect means of social control.[36]

Allee, an important figure in the history of ecology based on the community concept, and a Quaker sympathizer, took strong exception to the Norwegian's narrow sense of the foundation of biological order. Allee's name is correctly bound up with emphasis on the idea of cooperation as the most fundamental biological force. Disoperation, i.e., disorder, not competition, was the opposite of cooperation for Allee. The degree of common ground between the despot theorist and the Quaker is revealing; dominance need not mean a principle of autocratic rule. Dominance and subordination must rather be conceived as forms of social coordination. Like any other physiological object, dominance had to be understood in the full variety of its forms through careful prosecution of comparative biology. Allee's laboratory at Chicago was a scene of such comparative studies; a large part of the program was focused on bird flocks. But Allee studied organisms from bacteria, planaria, crustaceans, goldfish, and protozoa, to small mammals and birds in his exploration of the evolution of mechanisms of coordination from physiological mass action to differentiated nervous systems.[37]

## *The Community as Organism*

The community concept in ecology requires clarification before one can understand what Carpenter did on Barro Colorado Island in 1931. Allee is again the chief actor, and again the theme is the convergence of physiological perspectives from field and laboratory biologies before World War II. Allee defined the community as "a natural assemblage of organisms which, together with its habitat, has reached a survival level such that it is relatively independent of adjacent assemblages of equal rank; to this extent, given radiant energy, it is self-sustaining."[38] The self-sustaining community had all the other characteristics of an organism: development over time, differentiation of parts by form and function, organization by gradients into sub-wholes called fields, and above all, tendency toward homeostasis, "one of the major inclusive principles of ecology."[39] The community concept was initially developed by the generation before Allee, especially by the plant succession phytogeographer F. E. Clements and about 1913 by the animal ecologist under whom Allee earned his doctorate, Victor Shelford. Although the term *ecosystem* was coined in 1935, a cybernetic systems concept of the basic unit in ecology did not take form until after World War II. Primate studies took root in an ecology of physiological communities, for which the chief question was the coordinated action of parts to maintain an organismic whole.[40]

In his chapter on general physiology from 1900 to 1930, Garland Allen stresses the synthetic organic (materialist) holism that was the fruit of the work of men like Walter Cannon, Lawrence J. Henderson, and Charles Sherrington. These people worked out the concepts and techniques to make the terms *organization* and *integration* the everyday concrete material of twentieth-century physiology. Similar organismic concepts ruled embryology, or developmental physiology, in the late 1920s and 1930s.[41] In order to understand the depth to which the physiological functionalism of this biological framework was structured by the logic of dominance as the basic principle of order, it would be useful to follow the argument of two organicists important to Carpenter: Charles Manning Child and Alfred Earl Emerson.

For the developmental physiologist, Child, dominance initially meant merely rate of energy expenditure. But rates of exchange established dominance. Indeed, the detailed study of *rates* in biological systems was the principal motor of the transformation from physiological to cybernetic functionalism in developmental biology and ecology. Closely connected with the measure of rates was the measure of pattern maintenance or communication. For Child, the pattern of protoplasm, in all its forms from simple axiate organisms to complex society, was a behavior pattern determined by the "dominant region" of greatest activity in production and consumption of energy:

> Only the simplest sort of integration is possible without definite and more or less persistent dominance, that is leadership. . . . Apparently all that is necessary for the beginning of orderly integration in protoplasm is a quantitative difference in rate of living and the possibility of communication. Dominance or leadership in its most general physiological form apparently originates in the more rapid liberation of energy.[42]

Organization without dominance had to remain simple, exemplified by communistic zooids of colonial organisms. If the dominant or active region of an organism were removed—the classic logic of the physiological defect experiment—a new center of dominant activity must be established or organization would not reappear. Higher behavioral patterns depended on differentiation of parts and specialization of function. By no means did the greatest differentiation result from the most autocratic organization; effective control producing greatest freedom (homeostatis, in Cannon's language) was the fruit of the most complex development of communication of materials and processing of energy:

> Analysis of these integrative relations in organisms and society show that they are relations of some degree of dominance and subordination, of control and being controlled. . . . Without the dominant region the piece cannot give rise to anything above the body level from which it was taken; it possesses no initiative and remains only a piece. . . . [The] more completely communistic organism complexes, consisting of many similar zooids, have remained relatively primitive in

character, the chief relation between the components being nutritive. . . . As in the organism [in social integration], with the development of means of communication and transport, the effective range of social dominance increases greatly.[43]

Child's work remained controversial in developmental physiology and psychobiology; but the point of controversy was not the principle of dominance as a principle of coordination, but specific experimental verifications and the extent to which the individual and social isomorphism held. On the latter point, the controversy was about the validity of the theory of the superorganism. Based on his studies of insect societies, Emerson was a principal advocate of the superorganism, with the corollary that homeostasis was the correct term to denote social integration. Discovering the laws of dynamic equilibrium, i.e., discovering the organic variation and regularity of patterns maintained by dominance, was the task of the biologist in the laboratory and in the field.[44]

This section began in Stone's laboratory, at Stanford, where comparative psychology and reproductive physiology reigned. To understand the meaning of these functionalist disciplines, which Carpenter brought with him into the field to study monkeys and apes, subsections briefly considered the importance of dominance as a coordinating principle in physiology—whether observed in ecology, population biology, general physiology, development and embryology, or animal sociology. Far from blocking a science facilitating social peace and cooperation, the principle of dominance seemed to open up an understanding of the laws and therapeutics of harmonious social wholes. Dominance *could* mean extreme hierarchy, but that was only one of its forms. Higher life depended on more complex systems of dominance, including cooperation. Borrowing heavily from the work of people who had studied birds and insects, as well as small mammals, in the field and laboratory, Carpenter was prepared to contribute to a physiology of primate society, to a science of social coordination and control through dominance. This discussion began with Carpenter's unexpected chance to study wild New World monkeys in Central America as part of Yerkes's program of experimental sociology and with his rich inheritance from ornithology. Let us return to 1931 and Chapman's suggestion.

### *The Howlers of Barro Colorado*

On October 1, 1931, Yerkes wrote Carpenter about the opportunity "for naturalistic study of social relations, and particularly of familial and sexual behavior, of monkeys. . . . I agree with Doctor Chapman that the opportunity probably is unique and ever since he first suggested it to me, some two years ago, I have been looking and waiting for [the] suitable person to undertake the job."[45] On December 17, 1931, Carpenter was on a United Fruit Company boat bound for Panama, supported by a grant to Yerkes from the Committee for Research in Problems of Sex and by a National Research Fellowship.[46] Carpenter's first enthusiastic letter back to Yerkes announced the expected themes: he looked immediately for sexual behavior and for social bonds—expressed by "social skin treatment."[47] And in close contact with Chapman, he began the census of howlers on the island, counting first groups close to the laboratory. Chapman wrote Yerkes to express his great pleasure in the work: "The field is new, he has had largely to blaze his own trail. . . . I recall no census of its kind in Continental America. It is in itself a notable piece of work, but is, of course, only a beginning. There remain a study of the interrelation of clans and of life within the clan."[48]

Carpenter's early questions and records in the field made concrete the abstract logic of physiological functionalism of social groups:

> It is feasible to gather data on the following subjects: Territorial distribution; gregariousness and the types of intra-group consorts; progression; postures; manual, caudal, and pedal dexterity; food and feeding . . . ; sexual activity and reproduction (One instance of secondary and primary sexual behavior has been observed); maternal behavior; ontogenesis; play . . . ; fighting; defensive and protective behavior; vocalization, its frequency, time of occurrence and provocation; diseases and parasites; unified group actions and the situations in which they occur; relations of the males to the groups; the isolated males; specializations in the group; social facilitation; emotional expressions; reactions of the groups to the observer; instrumentation in arboreal progression . . . ; exploratory behavior; behavior suggestive of memory; "multiple Choice" situations; examples of delayed reactions; and instances of complicated behavior which may be considered evidences of "intelligence." I am attempting to work out the rhythms of daily activity as related to progression, rest, and feeding. The distances travelled are estimated. At the present time I am working on a scheme for scoring quantitatively the frequency of certain standard behavior patterns and the number of times that the different types of sub-groups, i.e., male and female, mother and infant, female and young, female with infant and young, and female with young and a male, are observed.[49]

Carpenter's list followed closely the plan Robert and Ada Yerkes used in *The Great Apes* in summing the extant knowledge of nonhuman primates.[50] The list highlights the concerns of comparative psychology and reproductive physiology.

Sexuality and its relation to group coordination was central. Robert Yerkes had sent Carpenter a copy of the new book by Solly Zuckerman, *The Social Life of Monkeys and Apes,* which argued that constant female sexual receptivity was the foundation of primate society and that dominance hierarchies and related fighting among males and male control of females to amass a docile harem were the mechanisms of social formation and maintenance in all primates. These arguments were based on observations of Hamadryas baboons in the London zoo.[51] Carpenter indeed found Zuckerman's suggestions pregnant. He collected animals (with special permission, since hunting was forbidden on Barro Colorado Island) "to determine the physiological state of the reproductive system of the females closely associated with males" and had "two very interesting experiments in progress. 1) I have selected a group of nine animals with one male. I plan to remove the male and to observe constantly the group during the period of reorganization of the group . . . [and] 2) The group that we have recently found consists of five animals . . . [I]t will be possible to completely isolate this clan, and since it contains about the average number of females per male, and since the animals can be identified individually, many important questions can be answered. . . ."[52] Following the logic of Child, the plan to remove the luckless male was the first of many such efforts to perform the classical defect experiment on the physiological social group. Collecting female reproductive tracts for anatomical study in order to answer questions about the relation, if any, of physiological state and social association continued throughout Carpenter's early work. Observations of deliberately isolated social groups, especially with reference to detailed study of pairwise relations of identified animals, was a basic technique for studying social bonds microscopically. We shall see later how important it was to Carpenter to determine the "average number of females per male"; a basic bionomic law was at stake. Later, Carpenter also collected males, either solitary or in all-male groups, to determine their age and sexual state in relation to their peculiar social grouping. It was hard to account for all-male *groups,* made up of males in good physicial condition, on the reigning hypothesis of heterosexual attraction as the core of organic society. A corollary hypothesis of saturation of available female sexual drives, thus producing supernumerary males, was predictable and was forthcoming.[53]

Carpenter's collection of reproductive tracts led him into relationship with two of the major reproductive anatomists of the period, George Wislocki, of Harvard, and Carl Hartman, of the Carnegie Institution's Department of Embryology, in Baltimore. To connect his behavioral records with precise physiological information, he needed expert help.[54] Wislocki, a member of the expedition to another area of Panama in June 1932 to study spider monkeys, taught Carpenter a great deal of sexual anatomy, in the course of trying to learn the duration of the reproductive cycle of a New World monkey.[55] Zuckerman's book and Wislocki's and Hartman's anatomical work were all part of the larger program to determine the meaning of primate female sexuality, especially the menstrual cycle, in the maintenance of societies. Whether or not primates had annual breeding seasons was a subquestion within that framework.[56] Another was the measurement of duration of menstrual cycles. Hartman had determined the length of the cycle in an Old World macaque, the rhesus monkey.[57]

Encouraging continuance of fieldwork, Hartman and Wislocki both had subsequent influence on Carpenter's career.[58] Hartman was a referee for the grant application to the Markle Foundation in 1938 to fund the Cayo Santiago rhesus colony in Puerto Rico. He originally argued against funding because of the isolation of the colony from a major university medical school. Only the assurance of Columbia University officials that Columbia researchers would use the facility changed his mind: "Engle and Philip Smith of the [Columbia] College of Physicians and Surgeons became interested in extending their studies of pituitary hormones and other endocrine secretions to nonhuman primates, and also interested in conducting new programs of research on behavior, particularly sexual behavior, under semi-free ranging conditions for comparison with results obtained in the laboratory."[59] Hartman was also skeptical of the Yerkes Orange Park laboratories for similar reasons—lack of geographical integration with a major center of science. He turned down the directorship at Orange Park after Robert Yerkes's retirement. Hartman was right about underuse of Cayo Santiago from 1940 till the 1950s. He believed that only the behavioral biologists like Carpenter ever used the facility correctly; and they were not dominant in its affairs, even though Carpenter was important in its establishment. The whole affair of Cayo Santiago illustrates the limitations of the "workshop" organization of scientific activity.[60] But here its significance is the continuing interlocking of themes in reproductive physiology and primate studies at the level of determining the founding of laboratories before World War II.

Wislocki played a more visible role than Hartman in Carpenter's work after the first howler study. In addition to encouragement,[61] he sought Carpenter's 5- and 10-year written research plans with the intention of including the younger man in the long-range program in primate reproductive biology.[62] Wislocki helped Carpenter plan his part in the Asiatic Primate Expedition and solicited Cannon's favorable consideration of the A.P.E. proposal in the Committee for Research in Problems of Sex.[63]

Carpenter's field notes for the howler study show from the beginning his personal and theoretical foundations in reproductive physiology informed by dominance, the homeostasis of social groups, and comparative psychology of drives and learning. His fulfilled expectations and surprises are especially revealing: "It is surprising that the two dominant males are always so closely associated spacially [no indication in the notes of how dominance was determined]." "Is it possible that at certain periods females do associate themselves as consorts [on the problem of accounting for females associated mainly with each other]." "Leading males seemed to be the most aggressive, hence probably the dominant [no description of behaviors called aggressive]." "This may have been a case of lack of submissive following by the 'focus' of the group [females with young] and hence splitting off [on a failure of females to follow putative leaders as a mechanism of social fission]." "Posturing is a constantly changing and dynamic type of overt homeostasis. . . . Behavior is the dynamics of posture [in the context of showing social contexts of posture in play and copulation]!" "This would suggest that there is no sharp active dominance-submissive gradient in the howler groups; no animal fears and gives way to the leader, dominant in group activity alone [i.e., vocalizations and leading progressions] and not over individuals [in relation to accounting in terms of the dominance theory for a lack of observed dominance interactions in howler groups—few observations did not mean irrelevance of dominance but low slope of the organizing gradient]." "The degree of sexual drive is probably fundamental to males continuing with the group [on Carpenter's early opinions on why males left a group]."[64]

Subject headings in the tables of contents for the field notes reinforce the categories of reproductive physiology: intragroup consorts, sexual activities and reproduction, maternal behavior, the isolated male. These categories were closely related to those heading observations of group integration, internal differentiation, and homeostasis: dominance-submissive gradient, relation of males to group, intragroup communication, intergroup communication, specialization within the group.[65]

Having looked in some detail at the contexts of Carpenter's work and at his initial reactions to the field experience, it is time to turn to the product: the finished monograph on howler monkeys, published in 1934, which was the sole visible artifact to contemporaries. How does this monograph structure the primate story from the building blocks we have been examining?[66]

### *The Monograph*

First, the organization of the paper established a pattern maintained in all Carpenter's further publications. The order of topics reinforced his claim that his naturalistic studies yielded only careful generalizations after the fact of observation.[67] The constant order was: field procedures; postures and vocalizations; feeding; territory and nomadism (progression through space); organization of population; integration of social group; coordination of social group; conclusions. Territory, population organization, integration, and coordination will form the core of the following analysis.

Using field techniques adapted from ornithology to understand the range of howlers, Carpenter followed the movements of several distinct groups and marked their positions on a field map. He compared the maps of 1932 with those of 1933. Mapping movement allowed him to make judgments about group leadership, composition, and mechanisms of maintaining territory (e.g., vocal battles between adjoining groups). Goal items within a territory might be food trees; they changed from period to period. One interpretation Carpenter gave to territory was physiological: "Thus it is seen that the possession of territory is not static by a dynamic adaptation."[68] The second dominant interpretive theme was psychological. "There is a network of pathways connecting the goals and sub-goals. Apparently as the group moves over certain pathways which lead to preferred goals, the activity is facilitated and the pathway is learned. . . . When a clan moves toward the border of its territory and the pathways and goals become less and less well-known, there is much frustration, progression is slowed, and the group becomes reoriented toward the more familiar pathways and better known goals."[69] The languages of objective animal psychology, establishment of neural pathways in the brain, and movement in external space are telescoped into a single set of terms and relations. Carpenter mapped the physical-psychological space of 23 groups, one in detail. The psychophysiology of space and movement was a basic component of physiological functionalism of primate groups.

Counting and identifying animals of age-sex class during progression brought Carpenter to the section on organization of a howler population. The census was the central operation and was intended to produce knowledge about group stability and composition; i.e., the census was related to a comparative sociology, not to questions in population genetics and evolutionary theory.[70] Therefore, the fundamental concepts Carpenter advanced were the bionomic principles of (1) *central grouping tendency* and (2) *socionomic sex ratio.* The central grouping tendency or "the mean tendency of gregariousness" was expressed in a formula for the proportion of animals of different categories characteristic of that species. Carpenter considered the tendency to be the result of fundamental physiological and psychobiological forces, not of ecological conditions. Ecological crises could disrupt the central grouping pattern, but they did not account for it. Thus, once the tendency was known, predictions about balanced group compositions and numbers of groups in a given space could be made. The formula for howlers based on combined 1932 and 1933 census data was: $(M16.4) + (f27.3) + (m13.5) + (i1\ 4.1) + (i2\ 8.0) + (i3\ 7.1) + (j7.6) + (j2\ 10.5) + (j3\ 5.5) + (M_I X)$. Males, females without infants, mothers, infants, juveniles, and isolated males were all included. (Only adult males were designated with capital letters, whether or not they were in the group.) This formula was the foundation for the second basic concept—the socionomic sex ratio, expressing the number of adult males to adult females in a stable group. Proper balance of forces explained the ratio: how did male and female sexual drives of a given species balance each other? If a sex ratio temporarily existed which left such forces out of balance, the observer could predict group recruitment, fission, fighting, or some other sign of instability that could trigger a homeostatic regulating mechanism. In several papers group splitting was called apoblastosis, a term for cell-division. Homeostatic mechanisms applied to all kinds of organic systems. Pattern- and size-regulating mechanisms were perceived in the society (superorganism), as well as in the cell and organism. Subgroupings of animals could be studied within the framework of balanced forces structuring the whole group and of the basic organizing axis of heterosexual drives. These relatively subsidiary and temporary subgroups were dynamic, differentiated parts of the social whole. The observer mapped spatial relations of animals to each other. "It would seem that the strength of the bond between animals is indicated by their spatial relations and the duration of particular spacing."[71] The motives and incentives determining subgroupings were studied as mechanisms of group integration and coordination.

Again, physical space was translated into the language of physiological psychobiology in order to account for sociological phenomena. That logic informed a whole practice in contemporary human psychology and sociology, from the study of small industrial work groups by Harvard Business School researchers, to the anthropologists' study of small groups under the rubric of personality and culture theory, to the sociometric school in social psychology. But the further explication of the social physiology of small groups, across biological and human social disciplines, must wait for the next section of this paper. Here the crucial point is that psychobiological forces were considered as mechanisms of coordination and integration of the social group within the logic of physiological functionalism. Attention was on the *whole;* explanations to account for the whole were field theories relying on equilibration of opposing forces along axes of gradation of some parameter. Many subfields were integrated within the boundaries of the major field:

> The cohesiveness of a group is the result of the complex processes of group integration and group coordination. The term integration is used to mean those processes, including the native propensities [drives], which strongly and positively condition an individual to a number of other animals, and hence to a clan as a whole. By group coordination is meant the many stimulus acts and responses which relate the immediate activity of each individual in such a way as to result in unified group activity.[72]

Thus integration referred to processes occurring over the whole life span of individuals and groups, while coordination concerned a time slice of a community in terms of the interrelated forces of its immediate cohesion. The subject of integration was socialization; the subject of coordination was communication. Both were part of the theme of control.

Carpenter's main point in the section on integration was that control began with direct contact between infants and mothers and ended as highly symbolic control across space by posture, vocalization, gesture, and glance. The mechanism accounting for the change was the interaction of inherited tendencies and learning understood by conditioning theory. Carpenter practiced the normal science of the interactionist paradigm that resolved the heredity-environment controversies of the preceding period. Interactionism was invoked as much to account for complex animal behavior as for human doings. This point will be important when we come to the problem of the relations among natural and social sciences in the 1940s, at the end of the dominant period of physiological functionalism.

Carpenter's basic procedure for studying integration was to give detailed descriptions of elemental relations between selected pairings of organisms, such as *f-y, M-f, y-y, M-y,* etc. In the later paper, he used the formula [$N$ ($N-1$)/2] for the maximum number of possible pair-wise relations which would simultaneously constitute a social group. But not all pairs were equally important in equilibrating forces to establish a whole. The important ones had to do with sexually-defined interactions ordered by the dominance concept. "The male and female 'vectors' of motivation function reciprocally to bring the animals together and cause them to engage in inter-related patterns of satisfying behavior. . . . With repetition of the reproductive cycle in the female and with uninterrupted breeding throughout the year, the process of group integration through sexual behavior is repeatedly operative, establishing and reinforcing inter-sexual social bonds."[73] Other systems of pair-wise interactions were also important; overall, the field theory of social groups was grounded on mutual dependence analysis like that Henderson used to study stability of blood pH. Carpenter's conclusions for howlers emphasized the low slope of gradients of dominance in the integration of the groups. The need in a logic of dominance was not for particular mechanisms, like a linear dominance hierarchy, but for understanding of the precise forms of dominance operative in a given social group. Peaceful cooperation and communal activity could be easily accounted for by dominance theory.

Coordination concerned the mechanisms of social control considered synchronically. Gesture and vocalization were minutely described and recorded in the field on camera and tape. The framework was psychological, with attention to reciprocal conditioning and stimulus-response patterns effective in maintaining a whole social group. These early observations and analyses of Carpenter were the foundation of extension to primates of the biology of animal communication begun largely on birds. This biology became a cybernetics of communication in primatology after World War II. The field technology of the later period was very sophisticated compared to Carpenter's early cameras and recorders. The technology of communication, biological or cybernetic, corresponds to the mode of workshop or industrial production of the scientific product.

### *The Physiology of Dominance in Other Monkeys, College Men, and Married Couples*

Before moving to a consideration of semiotics and sociometry in the late 1930s in Carpenter's primatology, it would be useful to recapitulate the themes of a physiology of dominance rooted in sex by reviewing Carpenter's other planned or executed research before the Asiatic Primate Expedition. A.P.E. produced the gibbon monograph, where laws of language and laws of space were further developed as the basic signs of dominance.

At the end of his first year of postdoctoral study, Carpenter submitted a 3-year research proposal to the Guggenheim Memorial Foundation, a philanthropy committed to furthering sociological investigations. This unfunded proposal was the first of many disappointments for Carpenter in his attempts to find material support for ambitious primate study plans. It also represented very clearly the "workshop" organization in primatology, where an individual and an assistant (commonly, as in this case, scientist and educated wife), without an elaborate institutional context, constituted the basic system of research. The full cost of the research would have been $9,000. The plan was to conduct a comparative sociological study of several New World primate genera. The basic question was: Why do animals live in groups? The procedure, similar to the howler study, was to provide detailed life histories and an account of the "native matrix of ecological and sociological factors." The laboratory context of psychobiology was in sharp focus, and the field data were intended to provide background for the more basic laboratory discipline. The proposed step toward a "comparative sociology of primates" was expected to provide data, "if it is thought advisable, [which] may be interpreted and related to various institutions in human society."[74] And finally, the study of the howler genus *Allouatta* was to be followed by studies of *Saimiri, Aotus, Lagothrix,* and *Cebus,* with a specific hypothesis in mind, a hypothesis embedded in the widespread research program to determine the role of sexual physiology in complex primate societies: "After studying the howling monkeys, I shall continue with species which show a greater versatility of behavior; thus proceeding from the relatively simple to the more complex forms. . . . The definite periodicity of female sexual receptivity found in the howler may not be found in the cebus monkeys, whereas grooming, which is practically absent in the howling monkeys, seems to be a dominant type of psychobiological behavior among the species of Capuchins."[75]

The "communistic" behavior of howlers, with their low slopes of organizing dominance gradients, seemed to go along with a conception of their social order as more "simple" than that of the more "active" monkeys. The greater complexity might reasonably be expected to correlate with a different sexual physiology, specifically, more steadily receptive females providing incentives for constant social interaction and resulting dominance systems for maintaining social order.[76] This conception of role specialization as the mechanism of complex pattern formation and maintenance, here cast in terms of sexual division of labor, was common to both biological and social scientific analyses of wholes throughout this period. The notion was at least as old as Durkheim, if not Adam Smith, and has been constantly reworked according to the logic of physiological functionalism. The emphasis in this section has been on the physiology of dominance and Carpenter's roots in biological discourses like developmental biology, community ecology, and comparative psychobiology. But contemporary human science discourse was similarly structured, without falling into "biological reductionism." The social psychology and anthropology of small groups, with the family as the privileged object of analysis and manipulation, proceeded with the tools of role theory and specialization of function as the principal mechanisms of stability and equilibration in social structures of fields.[77] The next section will look closely at Carpenter's use of materials from social science, especially semiotics and sociometry, so as to show the essential isomorphism of biological and social scientific logic in the construction of theories of stable communications systems.[78] The refined developments of the physiology of homeostasis, for example at Harvard, and of the analysis of social communication, for example at Chicago, were both crucial to the retheorization of functionalism in technological and cybernetic terms during and after World War II.

Shortly before Carpenter accepted his first professional position, as a Lecturer at Bard College, Columbia University, in June 1934, he applied to do two pieces of research, one on two gorillas in the San Diego Zoo and the other in testing married couples in a study of marriage compatibility. Both of these were reasonable proposals from a person who had just completed a study in the sexual physiology of primate society. The Social Science Research Council provided a small sum for the study of sexual behavior of two immature gorillas, both of whom Carpenter discovered to be preadolescent males (contrary to previous diagnoses), during the summer of 1934. The most remarkable observation, heading the section in the published paper called "Sexual Activity, Postures, and Locomotion," was that there was no sexual behavior to observe that summer. The rest of the paper documented observations of the caged animals on concerns common to researchers trained in Yerkes's Yale Laboratories of Comparative Psychobiology, such as ability to learn, cooperative behavior, delayed reaction, insight, and temperament. The organizing interests were sex, dominance, and cooperation.[79]

Marriage compatibility was a central interest of the Committee for Research in Problems of Sex, which tried repeatedly without major success to establish a program of research in human sexual psychobiology during the 1920s and 1930s.[80] Lowell Kelley at Storrs, Connecticut, a recipient of funds from the committee for marital compatibility research, tested the validity of weights for predicting marital happiness, which were developed by Stanford's Lewis Terman.[81] In seeking to collaborate with Kelley, Carpenter was not moving conceptually far from his howler study. "I have been interested in the problem since 1928, from the point of view of motivation . . . Almost all of my experimental and observational work has related to the sexual behavior of animals, and more especially mating relationships. After about six years of study, I am not without some knowledge of basic motivations of social relations in the animal world."[82]

But Carpenter obtained a faculty position at Bard and did not need to prolong his days as a research associate. He tried to continue his primate studies, and got a small grant for a 2-month recensus of howlers on Barro Colorado Island in December 1935–January 1936. He took 1,200 feet of moving pictures on that trip. The main theoretical goal was to test the predictive success of the "mean tendency of gregariousness" and "socionomic sex ratio."[83] The principles were confirmed.

His other research question at Bard, though, was the problem of human cooperation. There was no conceptual discontinuity; studies of cooperation and social cohesion were to be pursued at all levels. Carpenter proposed to study the factors producing "various degrees of adjustment" among college men in dormitories. He planned measures along an axis from compatibility to antagonism. The psychobiological model rooted in physiological functionalism was identical with the one guiding nonhuman primate research. Based on measures of positive and negative "valences" between pairs of individuals, Carpenter proposed to derive a formula for a "compatibility index." The index could be used to predict compatibility of persons in small social groups. "An increased knowledge of the factors of social interaction will surely lead to a more complete understanding of the processes of social control and will make possible a greater degree of planned control."[84]

Carpenter was turned down, but happily confronted another major opportunity, the one that resulted in the full development of sociometery and semiotics of nonhuman primate society. Harold Coolidge, of Harvard's Museum of Comparative Zoology, was assembling the money and people for a major primate research expedition to Asia; Carpenter would have the opportunity to study the gibbon, a form more closely related to human beings than the lowly howler. The gibbon was not only an ape; it had an upright posture and a monogamous sexual life.

### *Recapitulation*

Carpenter began his studies of the physiology of primate social groups with a question dictated by his most basic assumptions: Why do animals live in groups? If the animal physiologist needed ultimately to explain the coordination of the whole organism, the student of animal society needed to account for "whole animals in *complete groups* and these in their natural, complex and dynamic environment."[85] That which needed explanation was the *whole,* i.e., the phenomenon or organic organization. The tools for the explanation were developed within a framework of comparative, functional biology practiced in ecology, physiology, psychobiology, and zoology (ornithology). Carpenter drew from all of these sources to tell the primate story.

Integration, coordination, control, communication: these were the central conceptual realities and practical goals within physiological functionalism. In that framework, Carpenter helped developed a physiology of dominance, or control, rooted in a bionomics of sex. His major concepts were the mean grouping tendency, socionomic sex ratio, social homeostatic mechanisms, fields, and axes of organization identified by activity gradients. The small organized group, like the howler monkey clans on Barro Colorado, not the neo-Darwinian population and gene pool, was the chief object of analysis. The whole group was theorized to result from an equilibrated balance of opposing forces, in which sexual attraction and repulsion played the leading role.

Interaction of elements became social role theory within the examination of the natural social group. The primacy of the small group, organicism, balanced social relations, and specialization of function as the mechanism of growth in stability were all elements shared by biological and social scientists throughout the 1930s. Carpenter's easy passage from primate studies to proposals to examine adjustment of married couples and college men was not exceptional, nor did it signal a failure to discard discredited biological reductionism. Indeed, as the next section argues, the decisive theoretical tools for the construction of primate groups derived from positivist philosophical linguistics (semiotics) and sociology (sociometry), at least as much from the physiological discourses examined in this section.

Let us follow Carpenter to view the gibbons of Thailand and orangutans of Sumatra, in order to understand the connections of the physiology of dominance with theories of communication in language and space in the late 1930s. These sciences of communication from both physiology and linguistics were crucial to the physicalist cybernetic functionalism adopted in many areas in biology after World War II. The next section will begin with the Asiatic Primate Expedition of 1937, continue with the story of Cayo Santiago, and conclude with Carpenter's participation in two revealing conference volumes, one from a joint social and biological sciences' celebration of the 50th anniversary of the founding of the University of Chicago and the other from a collection of social scientific and biological papers on sociometry.

## 3. Semiotics and Sociometry: Laws of Language, Laws of Space, 1936 through World War II

> Men are the dominant sign-using animals.
>
> —Charles Morris, 1938

> I am fully aware that sociometry might have come into existence without me, just as sociology would have come into existence in France without Comte, and Marxism in Germany and Russia without Marx.
>
> —J. L. Moreno, 1960

There were many similarities between Harold Coolidge and Frank Chapman, men in a position to further Ray Carpenter's studies of wild primates. Without doctorates, both were consummate specialists in a zoological field. Both were important in the professionalism of their specialties, Chapman from the American Museum of Natural History Bird Department, Coolidge from the mammalogy section of the Museum of Comparative Zoology at Harvard. And finally, both were deeply interested and instrumental in effecting conservation of wildlife nationally and internationally. Chapman had been the occasion not only of Carpenter's opportunity to study howlers, but of his approach to the task. An assistant curator of mammals and secretary of the American Committee of the International Wildlife Protection Society, Coolidge played a similar role in the gibbon and orangutan research.

Legally enforced conservation was a mixed blessing to anatomists and physiologists; it made collecting much more difficult, if it went some way to making sure there would be something to collect. Taxidermist Carl Akeley, famous for his role in establishing the gorilla sanctuary in Parc Albert in the Belgian Congo just after World War I, had made the difficult transition to "hunting with a camera" from the more usual safari.[86] In the 1930s, scientists and travelers were only beginning to adjust to the individual consequences of rational conservation. Coolidge selected Asia for the primate expedition because the London Convention on Africa, which went into effect in January 1936, strictly protected the gorilla and partly protected the chimpanzee:

> It is true that specimens can be obtained on a scientific permit, but this new law is so recent that it will be some years before what one might call a liberal scientific permit would be issued. In Asia the Dutch have already closed down on the gibbon and the orang. The British are planning a Pan-Asiatic Convention similar to the African Convention. . . . When this happens, it will be too late to obtain a sufficient series of one species to make much needed studies of variation. It will

be also extremely difficult to obtain reproductive material and especially fetuses, which are at present extremely uncommon in the collections. . . . In years to come we can continue our field studies on the gibbon, but in order to obtain collections to correlate with these field studies it will be necessary to have a scientific permit, and such a permit would not be extensive enough to justify the expense of taking a party of primate specialists into the field.[87]

Steps to maintain scientific dominance in access to material have been materially important in the history of primatology. The ape expedition, then, headed for Asia.

The Asiatic Primate Expedition illustrated cooperative scientific organization, going beyond the single investigator in the field, required for expensive research.[88] Carpenter, of Columbia University, would study social relations. Wislocki, of Harvard Medical School, originally hoped to go to study reproductive anatomy, but could not. Harold Coolidge and Adolph Schultz, physical anthropologist from The Johns Hopkins University Medical School, with a young Harvard graduate student (Sherwood Washburn) as assistant, were the principal anatomical and taxonomic experts. The expedition was completed by John Coolidge as artist and photographer; J. A. Griswold, a Harvard research assistant in the Museum of Comparative Zoology; and Andrew Wylie, a Washington, D.C., special assistant for collecting large mammals. Wislocki, Robert Yerkes, and Walter Miles sponsored Carpenter. C. J. Warden, psychologist at Columbia, and A. T. Poffenberger, of the Columbia Council for Research in Social Science, presented Carpenter's plans to the Columbia Council to get funding. Carpenter's field notes listed 39 individual financial sponsors of the expedition. The list included the family names Barbour, Coolidge, Wheeler, Stone, Emerson, Shattuck, Terman, Miles, Yerkes, Hartman, Merriam, Frick, Cabot, Wislocki, Lashley, and Warden.[89] Social organization, popular interest, personal money, and scientific expertise from areas of psychology, anthropology, and reproductive and neural physiology were all behind the important expedition. The party sailed in late December of 1936.

Carpenter was in the field in Siam from February to June 1937. Two months and 14,000 miles from the U.S. a discouraged Carpenter recorded in his field notes that he had still not seen a gibbon, despite hearing them constantly. The theme of the difficulty of observing primates has recurred constantly in the primate literature. Compounding Carpenter's difficulties was the extensive collecting going on too near his study area. A move of camp produced opportunity for observations underlying the "Field Study in Siam of the Behavior and Social Relations of the Gibbon *(Hylobates lar)*."[90] The "viewpoint" section of the paper clarified the evolutionary framework, as well as the important point of the relation of this sort of study to the human sciences. First, behavior must be studied just like a problem in comparative morphology. Then, "it is sufficient to assume that just as there are structural relationships which place man in the same categories with the more complexly developed primates, so there are basic human needs, drives and types of behavior which have elements in common with similar functions of the non-human primate level. For example, many aspects of sexual behavior are similar in man and the apes. Perhaps in these primates one may observe *anlagen* [embryological term for primordia] of human motivation and behavior, free from cultural veneers and far enough removed to avoid the well known errors involved in man's study of himself."[91] Carpenter was sensitive to the mistake of describing analogous behavior as homologous: he intended to avoid both anthropomorphism and "theriomorphism"—falsely attributing animal characteristics to human beings.

### *Primate Anlagen and the Culture Concept*

His program was not unacceptable to the cultural anthropologists of the Boasian camp; rather it was directly consistent with the rationale for primate studies developed by one of its major spokespersons, Clark Wissler, of Yale's Institute of Psychology and the American Museum of Natural History. In fact, Carpenter's claim represented the position that emerged *after* it was unacceptable to study "primitive" people as simple windows into "civilized" behavior. The nature-culture distinction of social science, which was ascendant in the 1930s, did not deny the organic nature of human beings, but asserted the cultural, language-based control of organic raw materials, especially sex, by means of unique human mechanisms, like kinship, within socially determined systems, called cultures.

Wissler had been a grantee of the Committee for Research in Problems of Sex from 1928–33 as part of the committee's effort to establish a program in human psychobiology of sex. Sophie Aberle and Beatrice Blackwood actually conducted the research, among American Indians and Solomon Islanders, which was intended to probe simpler reproductive relations that might be useful in treating the sexual pathologies found among more complicated civilized people.[92] In 1933 Wissler reflected on this abandoned framework:

> The reasons why a program among primitive peoples was considered promising may be formulated as follows: 1) The assumption has been made that civilization is artificial as opposed to the natural. In keeping with this assumption, it is often said that the artificial settings of civilized society interfere with normal sex life. Accepting this as a working hypothesis, the procedure would be to gather information on sex life as observable among the uncivilized. 2) Experience eloquently preaches the necessity of comparative methods when confronted with problems of behavior and morphology. Hence it was assumed that the contrasts in pattern between uncivilized and civilized would assist in clarifying the patterns of modern sex behavior.[93]

*All* humans had "artificial" culture; only primates could show what once was sought among people. The culture concept did not challenge basic levels of biologization and sexualization in the explanation of society, but it did displace the position at which one expected to see the operation of drives unobscured by culture and self-consciousness:

> Now it appears that the surviving so-called primitive peoples, also, present artificial settings comparable to those observable in civilization; they exercise social control in forms similar to those exercised by civilized peoples. Hence no important research lead is apparent. Turning next to the comparative approach we note that the intensive study of chimpanzee behavior has been encouraged by the Committee. The preliminary results in hand suggest that the patterns of response for these non-human creatures will lead to a formulation of the human pattern.[94]

Two directions for sex research relevant to people were reasonable: comparative work restricted to primates, not primitives, and direct sexual studies of civilized humans, like those conducted by Kinsey a little later: "One set of behavior patterns seems to prevail throughout mankind, and so can be better studied among ourselves. . . . In the main, the sex behavior pattern and the integration of the sexes in group life can be clarified by comparing the anthropoids and man. Hence, the intensive study of the great apes is to be favored as the immediately important step."[95]

It is important not to misjudge Carpenter's point of view as old-fashioned biological reductionism. Otherwise, the history of primatology since 1930 becomes inexplicable. Wissler was only one spokesperson, but we will see later that Carpenter's framework could appear comfortably with those of prominent social scientists of the ascendant culture school: Alfred L. Kroeber, M. F. Ashley Montagu, and Robert Park. Controversies certainly remained, as around the legitimacy of the superorganism concept in human science. But primates had a very important role to play in post-Spencerian American social science. In 1936 Wissler reinforced his position on the relevance of comparative primatology for human sciences. He wrote to President Angell of Yale University in the context of a proposed reorganization of Robert Yerkes's research station:

> My hope at the outset [of the Yale Laboratories of Comparative Psychobiology] was that the gorilla and chimpanzee could be used in experimental approaches to social and cultural problems. . . . The behavior studies so far have indirectly influenced problem formulation in certain sectors of sociology and anthropology to the end that a few mature investigators frequently discuss informally possible problem set-ups using such primate material. . . . The outcome rests wholly in the availability of primates and the maturing of two or three young persons. . . . A growing group of anthropologists is seriously concerned with personality problems and seeking ways of handling such problems among primitive peoples. This is but a phase of personality study in general, which, in its larger aspects, promises to be the future area in which psychiatry, sociology and anthropology will attempt to work concentrically. Here again the availability of primates will be an important consideration.[96]

In the changed theoretical and material conditions of post-World War II America, Wissler's position remained relevant.

*Organicism*

In addition to arguing the place of primatology in human science and the continuing privileged place of sex studies, Carpenter was clear about the theoretical framework of organicism appropriate to the relation of biological and social sciences in the 1930s. Holistic field theories that allowed many interacting variables and connection of levels of organization without reductionism were demanded. For Carpenter that meant, "this study has not only dealt with descriptions of 'animals as wholes' but with 'whole animals' in *complete groups* and these in their natural, complex and dynamic environment."[97] This statement was consistent with the technique of microscopic observation of elemental pair-wise relations of known individuals, just as organicism was consistent with physiological investigations of subsystems. "Mechanism" in that sense was utterly uncontroversial. In fact, shared organicism characterized the work of Sherrington, Cannon, Henderson, and Carpenter; and they were typical of biological work from a physiological functionalist standpoint. The "nomogram" of Henderson charted the mutual dependence of pair-wise interactions in a whole complex buffer system of the blood; the "sociogram" of Carpenter did the same for social relations in homeostatically controlled primate groups. Henderson extended his laboratory physiology into sociology, via a reading of Vilfredo Pareto; and the social systems theorist Talcott Parsons knew his debt.[98] Carpenter imported into a biological study the sociological techniques of sociometry. The direction of borrowing was immaterial; the point is that, precisely at the period that Hamilton Cravens and other scholars cite as the end of creditable biological reductionism in American human science, both biological and social disciplines shared a logic that elaborated functionalist field theories. These field theories were the material directly transformed by cybernetic functionalism during and after World War II. It is at this point that the critical role of the physical sciences in midtwentieth-century biological and social science makes sense in the production of a unified science of control systems. Let us look further at this pattern in Carpenter's work.

Posture was the first level of organismic field analysis. "The orientation responses which are the products of the anatomical and physiological characteristics of the animal constitute a dynamic field which affects all of its behavioral patterns."[99] The locomotory-behavioral and individual psychological data (instrumentation, discrimination, temperament, etc.) were integrated with the next level, i.e., social interaction considered as a problem in field analysis. So followed the now familiar population census, description of sexes and ages of individuals in groups and as solitary animals, derivation of the socionomic sex ratio and central grouping tendency, and collection of whole groups to confirm observations of living animals. For gibbons the mean grouping pattern was a family: female, male, and their young. Solitary animals were explained as a consequence of the socionomic sex ratio for the species; sex was a basic organizing force.

> Three characteristics of reproductive behavior in gibbons relate importantly to their grouping patterns and more specifically to the male female relationship. 1) Apparently there is no definite breeding season, hence copulation may reinforce the male-female bond throughout the year. 2) As compared with macaques or chimpanzees, gibbons seem to have a low degree of sexual drive. 3) . . . [C]opulation may take place throughout the menstrual cycle and *even during pregnancy* . . . . These conditions . . . would indicate an equilibrated satisfaction of sexual hunger. . . . This kind of receptivity would seem to compensate for the lack of more than one female for each male and support the family pattern of grouping.[100]

Once the general field laws structuring the organic group were derived, so-called "microscopic analysis" of elemental paired relationships (such as male-female, male-young, etc.) was appropriate. Summation of such pairs gave a picture of group structure. All of this was precisely like the procedure in the howler monkey monograph. The basic strategy was to build a hierarchical organic structure integrating physiological (neural and endocrine), psychological (learning and drive), and social (integration and coordination) levels. The important difference distinguishing the gibbon study was the degree of detailed use of sociometric and semiotic analysis to explain an integrated control system. Both of these theoretical points of view were borrowed from human sciences. They were keys to the role of primatology in bridging the natural and social sciences in the midtwentieth century.

### *Sociometry*

Sociometric analysis was common in social psychology in the 1930s. Its basic tool was construction of the sociogram. In the words of the man who considered himself the founder of the sociometric movement, J. L. Moreno, "The proper placement of every individual and of all interrelations of individuals can be shown on a sociogram. It is at present the only available scheme which makes the dynamic structure of relationships within a group plain and which permits its concrete structural analysis."[101] The purpose of sociometry was expressed as facilitating the "self-realization" of group goals. "Maximum spontaneous participation" was seen as the only reliable means to achieve human social control. Social control and social structure were intimately linked concepts; the technical expression of this was the integration of the participant-observer into the group studied. In the internalization of control, the technique was maximally invasive. The fully integrated participant-observer, the only person who *knew* the group structure, could help the group to its end. Sociometry was essentially a "microscopic analysis" from inside the group. Thus it was only compatible with seeing sociology as analysis of small-group structures. For Moreno, sociometry revealed *motives,* "the psychological geography of a community."[102] The focus was not the individual, but the atomic social *relation.* Constructing boundaries of the group as a whole resulted from microscopic knowledge of parts studied in functional-structural relation. "The nucleus of relations is the smallest *social* structure in a community, the social atom."[103] A geometry of social relations allowed determination of the "tele" of a group—the goal around which it is *actually* organized (no matter how people might *think* it is organized), called the "group criterion." Once the criterion was known, the investigator could predict future group states and strategies effective in achieving goals—or in thwarting them. That is, sociometry included essentially "therapeutic and political procedures, aiming to aid individuals or groups to better adjustment."[104]

Typically, sociometry was a science producing control techniques. The essential problem was motion, action, behavior. "Sociometry is defined by its operations. . . . Sociometry is recognized by what it does, stirring to action and keeping action open but using scientific precision and experimental methods to keep action in bounds."[105] Carpenter's adoption of sociometric technique was consistent with his approach to building primatology as a science of control.

Group criteria differed in complexity; that was the fundamental distinction between animal and human organization. The sociometric expression for the difference was depth in a solid geometric volume. Animal sociometry could be adequately accomplished on surfaces, because relationships were unobscured by depth-producing factors like self-consciousness and language. Therefore, the problem of interference by the observer was simpler. So a sociogram for an animal society would be a two-dimensional map of physical relations in space, with arrows to indicate vectors of movement. Time was an additional factor pictured by a series of sociograms. Correct valuation of vector forces of attraction or repulsion should allow prediction of physical distances at a given future time. The psychological and physical maps were fully congruent. The geometry of social structures was totally different from chance spatial arrangement; the sociogram showed the actual psychosocial network tending to the achievement of group function. For babies and primates (not for primitives), sociometry was simply charting movement in space through time. "At the earliest developmental level [the embryological anlage of Carpenter], physical and social structure of space are congruous."[106]

At the same symposium in 1945 at which Moreno explained the relation of human and animal sociometry, Montagu waxed enthusiastic over the technique for anthropology. Montagu, a major spokesperson for the new human sciences, which insisted on a unique human biology whose product was language and culture, saw sociometry as the realization of Bronislaw Malinowski's "method of cultural analysis." Sociometry was the technique for structural-functionalism. It would play the key role in the "study of man as a functioning unit in the social continuum." The small groups studied by the ethnographer were ideal:

> Such a sociometric mapping of a group would not only yield an invaluable account of the social psychology of the group, it would also provide a more accurate picture of the culture as a whole than is obtainable by the usual means of ethnographic investigation. . . . But whereas functionalism is to anthropology as physiology is to anatomy, sociometry is to functionalism as histology and biochemistry is [sic] to physiology.[107]

So Carpenter, far from illicitly reducing human society to primate levels, adapted a well-regarded social science technique to the appropriate level of complexity. He had used sociometry without naming it in the howler studies. In the gibbon work, the approach was all-pervasive. In 1945, he outlined explicitly what he had done in mapping the pattern of a group. First, he made scatter diagrams of the whole group of animals in space; then he made vector diagrams of relations between individuals and reconstructed a map showing the summed positive and negative valences of interaction. Basic data were observations of spacing and duration of interactions. Each relationship was given a final character by "organismic summation." The basic concept was a *social relation,* "the reciprocal interaction of the behavior of two or more individuals which stimulate and respond to each other. The behavior, in turn should be considered as expressions of individual motivation or psychological processes."[108] The maps revealed social control of a group. Control always turned out to be based on some degree of dominance, usually but not always by adult males. *Status* was a measurable control quantity directly derived from dominance. Primates, unlike people, did not have elaborate extragroup controls. The semiclosed primate groups were managed by the socionomic sex ratio, social bonding patterns, and psychological conditioning to territory. Intragroup differentiation was limited by the degree of elaboration of dominance axes (status hierarchies). Intergroup relations were restricted by competitive mechanisms like territory, defense, and the absence of language. Territory remained fundamentally a psychological concept for Carpenter; it concerned motivation.

Let us return to the gibbon study to probe further how sociometric analysis functioned to produce a science of organismic control, and so knowledge of effective control strategies for ends either internal to the group or selected by an external manager.[109] With Carpenter, the place to begin is the analysis of a gibbon sociogram: a scatter diagram of spatial relations of discrete gibbon families. The maps were produced in the analysis of group territoriality.

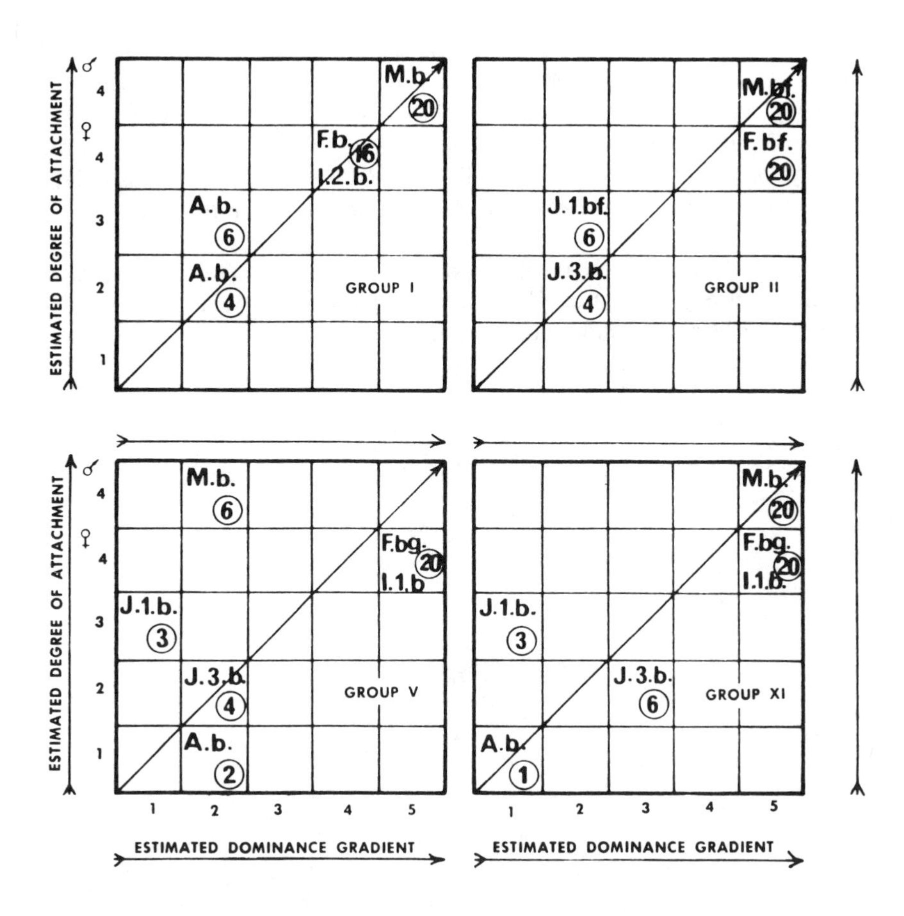


Figure 1. "Chart 2. Scatter diagrams of Groups 1, 2, 5, and 11. An attempt has been made to show two variables or axes within the group: a. Variation in dominance and b. Degree of attachment to the group, or a factor closely related to "conformity." Males and females are assigned equivalent attachment status. The estimated degree of dominance times the estimated degree of group attachment gives each animal a weighting in its group. A low rating, e.g., *Ab* in Group 11, indicates that the individual may soon leave the parental grouping. A high rating, e.g., Mb in Group 1, indicates for that animal a high degree of leadership status for group control. The spacing of the symbols which represent each individual in the group roughly indicates the observed degree of attachments among the apes. For example, the two young adults, *Ab* and *Ab* in Group 1, are strongly attached while they are less closely related to the adults of the same group." [From C.R. Carpenter, *Naturalistic Behavior of Nonhuman Primates* (University Park: Pennsylvania State University Press, 1964), p. 242.]

The horizontal axis of organization was the dominance gradient; the vertical axis was a gradient of attachment, which measured "conformity." Attachment multiplied by dominance position gave a measure of probability of an animal's joining or remaining in the group or migrating. The higher the "weight" of an animal, the more essential that individual was to group stability and control. Animals clustered together away from the heavy group focus might be the nucleus of a new group. The process of splitting was called mitosis or apoblastosis. Relations between groups were inferred by their spatial separation, maintained by actual fighting (rarely) or by symbolic interaction like vocalization and gesture (usually). A group was always heaviest, that is, more dominant, in the center of its own territory.

In his field notes, Carpenter drew several sociograms in an effort to predict future group states; his constant questions were, How do groups form? and How are they maintained?

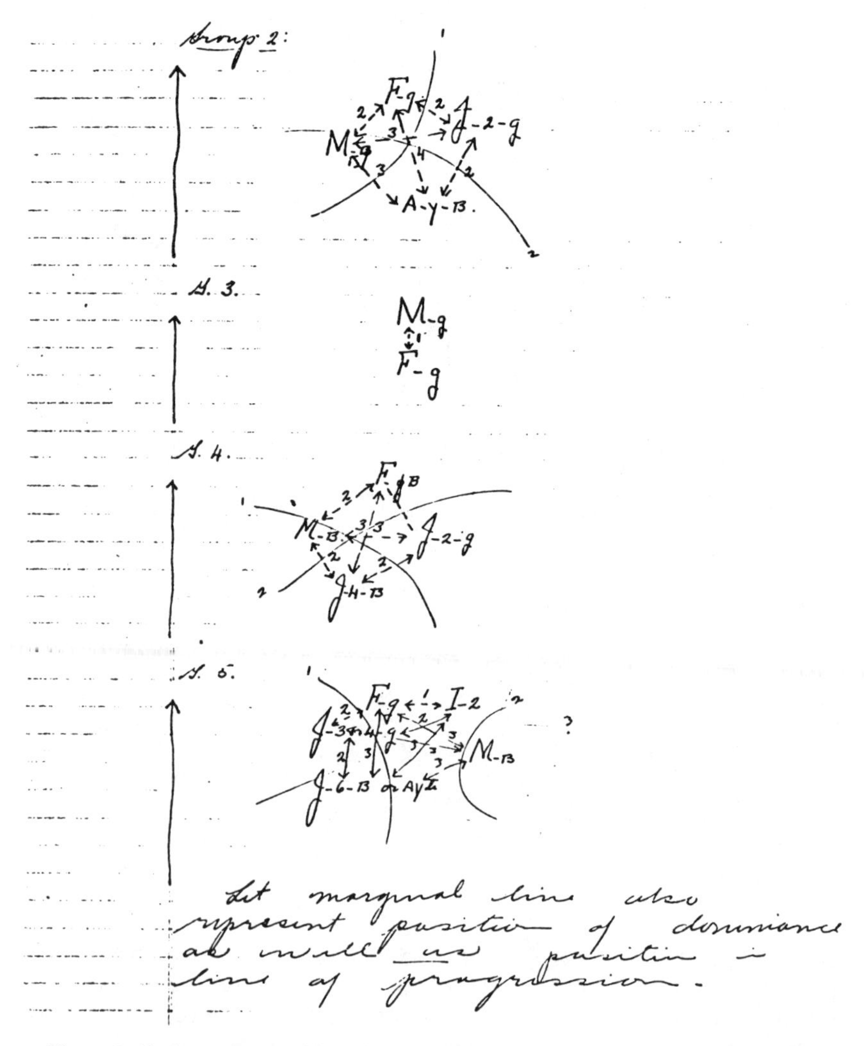


Figure 2. Sociometric mapping from Carpenter's field notes of the gibbon expedition. From C.R. Carpenter, gibbon field notes, p. 226, Carpenter Papers, in the special collections of the library of Pennsylvania State University.

The corollary of these questions was How might groups be scientifically managed? Note that in this scheme, behavior like fighting and competitive aggression, which establishes dominance gradients, was a mechanism of group integration. Appropriate amounts of such behavior were called leadership, control, and initiative. Inappropriate amounts were insufficient to hold a group together or were excessive and pathological expressions, which also disrupted groups. Carpenter believed these sorts of sociometric analyses for nonhuman primate groups were appropriate pilot studies for human sociometry, where control was much more complex. The connections were carefully made between animal dominance hierarchies as regulators of social integration and pathological human personalities, like the much-studied authoritarian character structure, with its dire social consequences.

*Semiotics*

Semiotics was the second borrowing from the human sciences important to primatology. Semiotics theorized communication as a problem in control systems. In its roots in the work of Charles Saunders Peirce, William James, James Dewey, and George Herbert Mead, semiotics was intimately intertwined with American pragmatism and behaviorism; a basic problem was to understand how systems of signs affected behavior patterns. Charles Morris, a philosopher at the University of Chicago, in *Foundations of the Theory of Signs,* defined semiotics as the science of signs, which studied things and properties of things in their functioning as signs. Morris believed that semiotics was the needed organon or instrument of all the sciences; it would be the tool of the unification of sciences in the twentieth century:

> The significance of semiotics as a science lies in the fact that it supplies the foundation for any special science of signs, such as linguistics, logic, mathematics. . . . The concept of sign may prove to be of importance in the unification of the social, psychological, and humanistic sciences in so far as these are distinguished from the physical and biological sciences. And since it will be shown that *signs are simply the objects studied by the biological and physical sciences related in certain complex functional processes,* any such unification of the formal sciences on the one hand, and the social, psychological, and humanistic sciences on the other, would provide the relevant material for the unification of these two sets of sciences with the physical and biological sciences.[110]

*Foundations* appeared in 1938 as the second number in the International Encyclopedia of Unified Science, whose editors were Otto Neurath, Rudolf Carnap, and Morris. Carpenter cited and used Morris's formulation as soon as it was published. Carpenter was nothing if not a positivist; semiotics met his need for theory in a way that the synthetic theory of evolution could not. Carpenter's constant goal was to produce a science of control.

Positivism and functionalism have important connections in the history of life and human sciences; these connections are critical to an understanding of Carpenter's version of the primate story. The founders of the school of urban sociology at the University of Chicago included Charles Horton Cooley, with his sociology of sympathy, and George Herbert Mead and Harry Stack Sullivan, with their psychological role theories. These approaches stressed interpersonal relations and processes of communication in the production of a functional social whole. Morris followed logically in this tradition in building a positivist, organicist theory of communication. When Carpenter systematically adopted these tools to dissect the primate social body, he naturally revealed a tissue of elemental social relations. The pathologies to which these tissues were subject—e.g., excess aggression, sexual malfunction, or disorganization of dominance relations—had to be understood as diseases of communication. Behavior, a dynamic physiological structure, was fundamentally the process of *signaling.* The organism produced signs, which permitted the interactions constituting or threatening the social whole. Therapeutics and interpretation of signs were never far apart. Life and human science have never strayed far from the clinic.

The content of semiotics was semiosis, or the process in which something functions as a sign. It was necessarily a relational study; i.e., ruled by a functionalist approach. Morris named the three parts of semiotics as semantics, syntactics, and pragmatics; he pointed to the three corresponding branches of comparative biology—anatomy, ecology, and physiology. "The properties of being a sign, a designatum, an interpreter, or an interpretant are relational properties which things take on by participating in the functional process of semiosis. Semiotics, then, is not concerned with the study of a particular kind of object, but with ordinary objects in so far (and only in so far) as they participate in semiosis."[111] Language was like any other organismic object studied functionally by positivists.

Ordinary human languages, called universal languages because they could potentially represent anything, were the richest systems of signs. But for that very reason, ordinary language was not always the most useful sign system. Certain purposes called forth special and restricted languages.[112] While the most complex focus of semiotics might be human linguistics, all organisms could be considered from the point of view of semiotics, i.e., from the point of view of an organism's response to sign vehicles. No hint of mentalistic language need or should enter such a science.

Since most, if not all, signs have as their interpreters living organisms, it is a sufficiently accurate characterization of pragmatics to say that it deals with the biotic aspects of semiosis, that is, with all the psychological, biological, and sociological phenomena which occur in the functioning of signs. . . . The interpreter of a sign is an organism; the interpretant is the habit of the organism to respond, because of the sign vehicle, to absent objects which are relevant to a present problematic situation as if they were present. . . . The response to things through the intermediary of signs is thus biologically a continuation of the same process in which the distance senses have taken precedence over the contact senses in the control of conduct in higher animal forms. . . . Considered from the point of view of pragmatics, a linguistic structure is a system of behavior.[113]

These principles have been important in the transition from physiology to cybernetics as the functional correlate of pragmatics in primatology considered as a science of communication-control systems. In the transition, the organism as responder to the sign vehicle lost its privileged position. The more powerful analysis of sign systems, cybernetics, dispensed with the need for a biotic component, just as pragmatics had dispensed with the requirement of a conscious being in analyzing and effecting control of behavior through representation. Consciousness was the first casualty of the modern assault on vitalism; the organism was the second. Physiology with its associated psychobiologies (including behaviorism) was the sufficient instrument for the first task. Even "insight" and "foresight" came to be studied *within* objective psychology; Carpenter presented a good example. Cybernetics was the organ needed to complete the second project: the removal of the organism as the necessary object of biology. Morris continued to be an important philosopher in the transformation of primatology from physiological to cybernetic functionalism.[114] This issue will be explored in more detail in the last section of this essay.

Here it is sufficient to point ahead by juxtaposing anthropologist-linguist Thomas Sebeok's 1967 rephrasing of Morris's principle that the interpreter of signs is nearly always an organism. Sebeok spoke at a conference on social communication among primates organized by Stuart Altmann. Altmann was one of the first students of animal society to use the label "sociobiologist" to designate the investigator of natural communication systems. Sociobiology has become the science of "zoosemiotics," or in Sebeok's words, the "coding of information in cybernetic control processes and the consequences that are imposed by this categorization while living animals function as input-output linking devices in a biological version of the traditional information-theory circuit with a transcoder added."[115]

Carpenter's continuing interest in primate societies in terms of communication must be seen in the context of the project of semiotics. Practically, the primate investigator studied gestures and vocalizations, as well as the development of symbolic social control systems from the more primitive contact control appropriate to the early infant. Sound recordings, films, taxonomic and functional classifications of calls and movements, charting of the ontogeny of communication: these were the appropriate artifacts Carpenter produced in his gibbon study.[116] Theoretically these artifacts were the data for constructing a picture of primate society as a dynamic, integrated communication-control system. In terms of the project of the unification of modern human, life, and physical sciences, a semiotic understanding of primate society helped link the many levels of objects studied by science together without reducing one level to another. Human language was only one kind of control system. The limitation of gibbon integration and coordination was a consequence of the limitations in the power of their systems of signs. "Expressed differently, it may be said that gibbons have capacities for making complex perceptual social responses and certainly responses to symbolic cues such as gestures or meaningful facial expressions, but they do not have capacities for employing more complex, generalized, 'higher level' symbols."[117] Nonhuman primate communication systems were enlightening to the student of human communication, but not in any simple reductionist sense.

Neither semiotics nor sociometry contradicted a physiological approach to organic wholes. All three were action-oriented; all focused on behavior, on motion, on function. Their respective origins in philosophic linguistics, social psychology, or biology did not make them mutually incompatible, but rather different windows onto the same problems of coordinated wholes understood in terms of control and regulation. The organismic field theories from biology; the mapping techniques for showing force equilibria and disequilibria in social systems; and the unification of physical, physiological, psychological, ecological, and social levels of organization in terms of sign systems were all a proper part of Carpenter's primatology. His product was a detailed analytical system for moving from physiology (e.g., endocrines, reflexes) to psychology (behavior, learning, drives) to sociology (integrated, coordinated group and breeding population). The strong ideological lessons—e.g., slow change permits adjustment, frustration produces aggression, dominance is a mechanism of cooperation—were all conveyed without any resort to conflating nature and culture or their proper sciences. Finally, semiotics, sociometry, and physiology were incorporated into a cybernetic approach to primatology after the 1940s. Then the project of constructing a powerful theoretical and technical organon for understanding and building complex organized wholes with many simultaneously interacting variables was achieved. Mechanism, space, and sign; physiology, sociometry, and semiotics: all were part of the new tool.

### *Cayo Santiago and the Practice of Primatology*

To conclude the analysis of physiology, sociometry, and semiotics in a prewar primatology, let us follow Carpenter through one of the important consequences of the Asiatic Primate Expedition: the establishment of a semi-free-ranging rhesus colony on 37-acre Cayo Santiago, off of Puerto Rico.[118] In 1938 Bard College failed financially, but Carpenter managed to stay at Columbia as an assistant professor in the Department of Anatomy of the College of Physicians and Surgeons. In 1937 Phillip Smith and Earl Engle, with Carpenter, planned a rhesus monkey colony to be located at Columbia's associated School of Tropical Medicine in the University of Puerto Rico. A convergence of interests from reproductive physiology, naturalistic behavior studies, and investigation of infectious diseases such as polio and tuberculosis was behind the successful application to the Markle Foundation for a 3-year grant to establish the colony. A partial Indian embargo on the export of rhesus monkeys was an additional stimulus.[119] Originally, Carpenter hoped to establish gibbons as well as rhesus on Cayo Santiago, but the apes proved too difficult to keep under the available conditions. The project to establish free-ranging colonies was seen as a necessary link in the chain of research facilities, from field stations to specialized laboratories, in order to produce several different kinds of primates "standardized for biological research."[120] On December 2, 1938, the approximately 450 Indian monkeys arrived in Puerto Rico, after a difficult 6-week ship passage. After testing them for tuberculosis, Carpenter released the animals onto the island. He would remain associated with the Cayo Santiago rhesus until he left Columbia in 1940, before the termination of the Markle grant. He would not return to Cayo Santiago for over 15 years, when primatology again caught medical and social interest.

Carpenter's plans and conduct on Cayo Santiago are an ideal window into the interconnections of physiological functionalism and sociometry-semiotics in the study of organized groups. After protecting against disease, the first priority in the field notes written on ship with the monkeys was, "Mates must be graded as to sexual potency and their rank order dominance established." The list continued with plans to study castrated animals, maternal behavior, structures of dominance before release, and brain lesions.[121] Typical problems Carpenter hoped to pursue were: "Select from the males to be released on the Island a number of individuals. Test them for sex drives and dominance by time sampling procedures and Murchison. Castrate them and then release them. Keep running records and test at intervals of three months."[122] "What are the sensory cues which serve as a basis for monkey females to discriminate their young infants?"[123] "Using pulling-in techniques—males pulling in females—What the communicative bases are for the selection—central vision, smell, hearing—"[124] "Produce experimental homosexuality."[125] "Produce intersexes by injections of internal secretions."[126] "Work on sex difference of dominance—Determine hierarchies for both sexes. Also species differences."[127] "Problem: Given animals with bi-lateral and uni-lateral frontal lobe-ectomies. To learn, what adaptations are made to a free-ranging environment and competitive social conditions."[128]

Time and money did not permit all of these problems to be pursued. The field notes from the period December 1938 to May 1940 revealed the priorities: (1) a study of dominance as the primary integrating mechanism of primate society, (2) sociometric mapping of dominance relations and other social bonds, and (3) analysis of inter- and intragroup interactions as signs in a functioning system. The basic technical practices were counting numbers and categories of animals in groups, construction of force-vector maps to show the logic of their positions and the probabilities of splits and fissures, and determination of the sexual status of females as a function of their significance in the bonding of males to the primate group. Females were bound to the group by the dominance of males; males were bound by the sexuality of females. Both were bound to each other by a logic of control. The product was the reproduction of primate society.[129]

Carpenter performed one 3-week "defect experiment" incorporating the physiological, sociometric, and semiotic principles of his primate story. He operated on the most dominant group that had formed on the island, the Diablo group with about 85 animals, including 7 adult males. He determined the order of dominance for the males only—the presumed axis of greatest activity in primate organization. Dominance was determined by observations such as, "When a group is motivated to travel but the Prime male does not move, the group makes little progress. Certain behavior cues from the prime male may set the group in rapid continuous motion. These cues consist of gross behavioral patterns which are directional."[130]

Carpenter believed he had devised a system for "determining dominance and social status items which will I believe differentiate the position of a group male after 20 to 25 hours of observation. The items include: sexual activity scores, intra group dominance and submissive scores and inter-group dominance scores."[131] With these determinations, experimental manipulations could begin. On June 4, Diablo was removed from the group and put in an outdoor cage nearby. For a week, Carpenter watched the group's behavior and territorial range. He witnessed confusion and disorganization, marked by increase in fights and decrease in space for the group's activities. He watched to see which of the remaining males would reestablish order, but for days saw only a "persisting lack of integration" and movement that was "amoeboid . . . sluggish and uncertain." [132] On June 11, he trapped and removed *M*174, the next male in the prestige hierarchy. On June 17, he removed the third most dominant male. The observed result was that previously subordinate males of the group occupied central positions among the females and young; but the group as a whole was in physiological decline: "As the male power of a group is reduced the boundaries of the group become more permeable."[133] He noted female fights during this period with special interest: "It would seem that these female fights relate to the restructuring of a group when a dominant male has been removed or displaced."[134] As in a hydroid polyp, the previously subordinate axes of organization became more visible when the dominant region was cut away. The spatial force-vector diagrams constructed before the trappings disintegrated as stable relations gave way to social flux. Finally, on June 26, Carpenter released all the captive males and recorded the reestablishment of social order. The analysis and construction of this defect experiment demanded the integration of tools and concepts from physiology, comparative psychology, social psychology, and neopositivist linguistics. Determination of endocrine status, spatial mapping with symbolic distances, recording gesture and movement in terms of their control value for the behavior of group members, and, above all, the basic questions of the study on the mechanisms of group foundation and maintenance required a synthetic view of primate society. Life and human sciences came together in the study of the anlage of human control.

### *The Collected Volumes: Celebration of Consensus*

The fiftieth anniversary celebration of the founding of the University of Chicago marked these relations of the life and human sciences which Carpenter helped to practice in the 1940s. The divisions of biological and social sciences came together to produce *Levels of Integration in Biological and Social Systems;* Carpenter's paper appeared physically midway between the contributions of Allee and Emerson, animal sociologists who based their work primarily on social insects and birds, and those of Kroeber and Park, an anthropologist and a sociologist identified with the triumph of Boasian social science in America.[135] The September 1942 symposium had two origins. First, the social science division desired to have a conference exploring the newly opened, promising research areas on the "borderland"[136] of the study of human society: (1) rapprochement of anthropology and sociology; (2) recent investigations of social behavior of monkeys and apes relevant to the origin of human society; and (3) work in mammalian and bird society that had caught the interest of sociologists and anthropologists. The borderlands could be explored now that the *boundaries* of life and human science seemed secure. Second, the biologists had planned papers exploring the problem of parts organized into wholes, from multicellularity to society. Clearly, the two planned sessions should be incorporated. The result was the celebratory volume signifying the relations of these sciences on the eve of America's entrance into World War II. The problems of integration were transformed by that war, in theory and in practice.

By August 1940 Carpenter had accepted a position in the Department of Education and Psychology at Pennsylvania State College. His relations with Smith, at Columbia, were strained, partly over the disposition of gibbons from Cayo Santiago, but largely due to different priorities in the use of the monkeys.[137] Carpenter's behavioral emphases were not central for Smith. Carpenter never had a stable position at Columbia, and he was not a powerful figure in the central research areas of reproductive and neural physiology. His behavioral studies were informed by reproductive physiology, but the interest was not entirely mutual. Most laboratory physiologists tended to look at primates in terms of supply of individual animals for experiments. Because most of the large funds were directed to establish laboratories, this bias plagued students of social

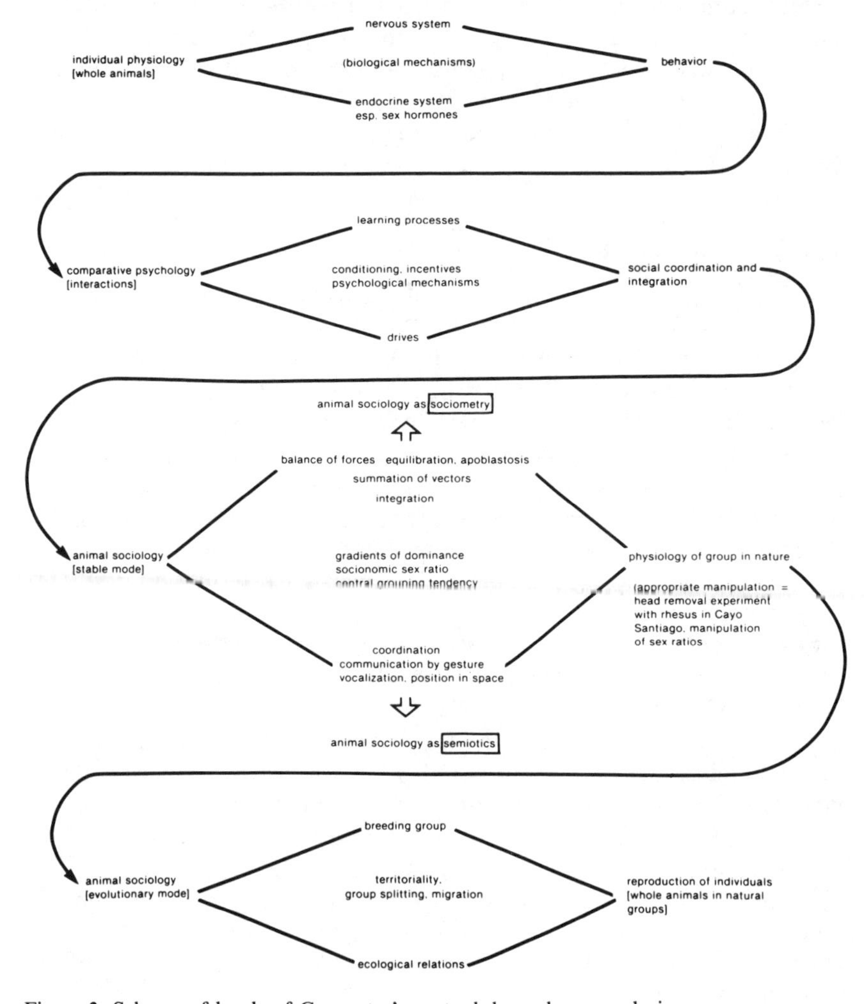


Figure 3. Scheme of levels of Carpenter's mutual dependence analysis.

behavior who did not have an independent research base. Carpenter's new location in a department of education and his departure from a medical school greatly affected his role in primate studies. Partly because of the war, partly because of widespread changes in research priorities, primatology was at a low ebb in the 1940s. Its rebirth on an international scale in the 1950s must be accounted for in terms of processes outside itself. Primatology's transformation was part of a powerful reworking of biology as a cybernetic engineering science.

## 4. A Cybernetics of Primate Society: Primate Nature as a Command-Control System, World War II through 1970

> Everywhere we met with a sympathetic hearing, and the vocabulary of the engineers soon became contaminated with the terms of the neurophysiologist.
>
> —Norbert Wiener, 1948

> In the systems under consideration here, consisting of a transmitter-transcoder-channel-receiver, the relaying mechanisms are members of the order Primates.
>
> —Thomas Sebeok, 1967

> Every object that biology studies is a system of systems. . . . The various levels of biological organization are united by the logic proper to reproduction. They are distinguished by the means of communication, the regulatory circuits and the internal logic proper to each system.
>
> —François Jacob, 1974

The principal distinction between physiological and cybernetic functionalism is the status of the concept of the organism and the surrounding logics of intervention. In cybernetic functionalism, the organism has lost its privileged position; it has given way to a more comprehensive, technical system, which has demanded a redistribution of relations among physical, life, and human science discourses. The functionalist logic of self-regulatory systems still pertains, but the communications engineer has replaced the sexual physiologist as the expert of choice. But the communications specialist, like the physiologist, remains occupied with the imperatives of systems reproduction. The construction and control of reproducing systems remains the key tool for the management and appropriation of the productive powers of organic-technical structures. Bionomics of sex merely gives way to systems management of communication. Thus sociobiology, a cybernetic science, has replaced psychobiology, a physiological discourse, in theorizing the fabric of social life. This section explores a fragment of this fundamental reordering of the structure of the objects of scientific knowledge.

Carpenter himself must disappear from the story for a time, in order to understand his acceptance of the reassembly of the elements of primate communication from the 1950s to his death. This section begins with an exploration of Carpenter's effort in 1947 to build a new educational machine, whose conception was made possible by wartime operations research. The connections of new ideologies of worldwide "interdependence" with psychiatry and education are briefly explored. Then the section sketches the roots of cybernetics in operations research conducted by men directly linked to the linguistic, physiological, and physical sciences of self-regulating systems. The section returns to the primate story with the work of Stuart Altmann and John Emlen, both major contributors to the sociobiological synthesis. The last part of the section returns to Carpenter to explore his use of the primate story (1) to manage war and human aggression, (2) to probe the design features of primate communication in parallel with his efforts to develop human telecommunications systems in the underdeveloped world, and (3) to understand his last fieldwork in collaboration with Rockland State Hospital and Yale University psychiatrists on free-ranging gibbons implanted with "telemetric devices intended to monitor aggressive and reproductive interactions."

### *Educational Technology*

During World War II, Carpenter served in the U.S. Air Force as technical adviser on training films for jungle warfare.[138] In 1944, still an air force captain, he was the chairman of the American Psychological Association's committee on audiovisual aids. In that context, he established the Psychological Film Registry at Pennsylvania State and directed the evaluation of films produced by the military for use in psychological instruction. He judged that the armed forces had made impressive progress in producing training aids which could be adapted for many educational purposes.[139] His interest in new educational technology continued after the war in his position as director of the Instructional Film Research Program at Pennsylvania State, and then in a variety of positions applying television as a tool of national and international integration. He was a reserve officer in the air force, with responsibilities in training technology, a special consultant on educational programs in Guam and Samoa, a participant in numerous international conferences on problems of integration through the use of new communications technologies, a consultant on legislation on multimedia approaches to primary and secondary education, a consultant to the General Electric Company and Raytheon EDEX Educational Systems, and a director of the Center for Applied Educational Research, in Washington. Under Ford Foundation auspices, he also advised the Indian government on development of a communications system. In short, Carpenter moved readily from the study of primate anlagen into the practice of human control.

Educational technology for Carpenter remained tied to a medical, psychological, and psychiatric approach to human cooperation. In the air force, he helped organize the German Youth Reeducation Program and was the Director of the Research Section of the Biarritz American University in France. He interpreted his work as part of rebuilding democracy through judicious application of educational control as a kind of psychiatric therapy to produce adjustment after pathology. "The German people have had the shock treatment. Now they need reeducation and readjustment. Who is to function as the psychatrist?"[140] The rhetorical question received an answer: the United States. Returning soldiers, defeated populations, and ordinary students all faced problems of readjustment to which a new technology could apply.

Carpenter's approach to his post-war educational mission had the highest sanctions. It is useful to look at them briefly in order to understand the roots of a persistent connection of postwar primatology with a psychiatry that addressed itself to mass social therapeutics—for example, in managing aggression and stress. This brief glimpse of the ideological ties of the promotion of mass mental health, education, and psychiatry will be useful later in this section, when Carpenter's collaboration with the psychiatrist José Delgado is explored in more detail.

C. B. Chisholm, a Canadian psychiatrist and former army officer who took a leading role in the World Health Organization, announced in 1946 the themes crucial to understanding Carpenter's rhetoric and practice immediately after the war. Chisholm identified educators and psychiatrists as key partners in a new mass preventive psychiatry addressed to a whole society with the same logic exercised previously on small groups. Again, the family was both a major concrete focus and an ideological model for therapeutics and theory. Educators and psychiatrists were to assume the burdens of parenthood in helping the world citizens adjust to the postwar realities of global integration and interdependence. More than ever, local prejudices would have to give place to a more inclusive identity prescribed by the experts. Therapeutic attitudes should replace religious outlooks in a psychiatrically ordered program of education. In the view of the former army psychiatrist who helped rationalize the mission of the World Health Organization, maturity meant the ability to work in organizations under rational authority and the willingness to show tolerance, adaptation, and compromise. Management of interpersonal relations was the task of educators and doctors.[141]

Carpenter's immersion in the ideology of promotion of international peace through education for integration informed his pursuit of educational technology. The control dimension of the new educational technology was not metaphorical. It was directly linked to developments that have come to be grouped under the term *cybernetics.* Let us move from a concrete piece of apparatus Carpenter wished to developed in 1947 to a glimpse of the material foundation of cybernetics in World War II. Then we may return to the primate story armed with new and forceful tools.

In 1947 Carpenter applied for a grant to the Viking Fund "For Developing and Constructing a Device for Facilitating Group Learning and Measurement of Results."[142] Carpenter saw the significance of the project in the adaptation of new electronic and computer technology to the immediate measurement of student learning in a lecture situation. The "proposed device system" would make it possible to measure learning without the usual delay imposed by traditional testing. Instead, each student would have a device at a "response station" on which to register answers to questions posed by the lecturer. The instructor would have a display panel that instantaneously registered the pattern of errors in the class. In particular, changes of attitude could be measured as the lecture was in progress. Engineers from the Navy Human Engineering Section would assist in building the device. The apparatus was conceived within the framework of operations research, which studied the human being within an overall system and measured error rates of both human and machine components in relation to the target, or desired behavioral responses.

Paul Fejos asked Robert Yerkes to evaluate Carpenter's proposal. Yerkes suggested that the director of the Viking Fund seek the opinions of "Vannevar Bush of the Carnegie Institution of Washington and of Professor Frederick E. Terman, engineer and specialist in electronics, and his father, Professor Lewis M. Terman, specialist in mental measurement, etc., both of Stanford."[143] These are the names which lead us directly to the story of the transformation from a physiological to a cybernetic approach to problems of integration in life and human science. It is fitting that the son of the psychobiologist Terman, who devoted his life to measurement of genius and gender adjustment within the framework of physiological functionalism, should appear in our story as an engineer in promotion of adaptation within the context of cybernetic functionalism. The story is much too large for the confines of this essay. Therefore, this work must restrict itself to the briefest suggestions that illuminate the reappearance of primate studies as a cybernetic communication science in the 1950s. Among the critical figures were Vannevar Bush, Norbert Wiener, Warren Weaver, and Arturo Rosenblueth.[144]

*Operations Research and Cybernetics*

During World War II Bush headed the Office of Scientific Research and Development (OSRD). The OSRD directed the efforts of over 6,000 scientists and engineers on the federal budget. The unprecedented coordinated material base was the foundation for the A-bomb, computer-controlled antiaircraft guns, radar defense systems, and industrial production of penicillin and antimalarial drugs. At the Massachusetts Institute of Technology before the war Bush had perfected the differential analyzer for regulating complex electronic distribution networks. His colleague at MIT, Wiener, who coined the word *cybernetics* around 1947, worked on the mathematics that helped refine Bush's machines. Also, in the 1930s Wiener participated in an informal seminar on scientific method at Harvard with Rosenblueth, a Mexican physiologist who collaborated with Cannon on study of the autonomic nervous system. Physicians, medical scientists, physicists, and mathematicians participated in this seminar, which like Henderson's seminars on Pareto, was an important social locus for working through the implications of a science of "organized complexity."[145] The organicism of Cannon and Henderson, which focused on the structure of self-regulating organic systems, was thoroughly incorporated into the fabric of cybernetic systems theory. Finally, in the 1930s, Weaver was redirecting the policies of the Division of Natural Sciences of the Rockefeller Foundation away from the physiological functionalist-psychobiology of men like Yerkes and toward the application of theory and techniques from physical science to biology as the foundation of the systems science of molecular biology. Weaver, a mathematical physicist who had held positions at the California Institute of Technology and the University of Wisconsin, was appointed by Bush during the war to head the Fire Control Section, responsible for gun control systems, under OSRD. Weaver was chief of the Applied Mathematics Panel of OSRD from 1940 to 1942. He was in close touch with the theoretical and practical developments of systems theory. Weaver published, in the same volume with Claude Shannon, of the Bell Telephone Laboratories, the paper on the historical background to Shannon's mathematical theory of information.[146] In sum, on the levels of connections of individual people, theoretical progress, technical capability, shared ideology, institutional bases, and large-scale historical possibility provided by war and depression, cybernetic functionalism was an organized complexity.

At the beginning of the war, Wiener wrote to Bush to argue for the utility of his prewar computer work. Bush wrote Weaver at the Fire Control Section on the matter, and Wiener was assigned with Julian Bigelow to study the problem of ideal gunnery control. Wiener looked for the machine analogue of the human gunpointer and airplane pilot, which operated by principles like negative feedback, known from servomechanisms. He sought the advice of the physiologist Rosenblueth. Together they considered the pattern of pathological breakdown in neurophysiological systems that produced greater and greater tremors or oscillations away from the desired goal. The collaboration led directly to the conception of the performance of the nervous system as an integrated whole with purposes, for which machine analogues could be conceived and built. They immediately saw that the work had philosophical as well as practical implications, resulting in the paper "Behavior, Purpose, and Teleology."[147] By 1944, Rosenblueth, Wiener, and their associates were explicit about the unity of a set of problems centering on communications, control, and statistical mechanics, whether in machine or living tissue. Wiener believed their work "made of communications engineering design a statistical science, a branch of statistical mechanics."[148] Control engineering and communications engineering became inseparable. In systems operating by stochastic processes, the problem was optimization of system design for adequate function under the strains and constraints of the machine-environment pair. The organism lost its privileged place in the study of organized complexity and became only one particularly interesting sort of self-regulating system. Operations research in Britain and America was predicated on this theoretical foundation, and the effects in biological sciences after the war were legion.[149]

Wiener summarized the themes of cybernetics in his popular book, *Cybernetics and Society: Human Use of Human Beings:* (1) the Gibbsean (statistical, thermodynamical) point of view has revolutionized modern life; (2) society can only be understood through the study of messages and communication facilities; and (3) physical functioning of the living individual and the operation of new communication machines are parallel in their attempts to control entropy through feedback.[150] These points of view were the ideological face of a transformation in the practice of science greatly advanced by wartime research: the industrialization of science. The processes and logics of control and automation comprehended under cybernetics have been part of a general retooling of functionalism, the science of integration and adaptation. Cybernetic theories and techniques have permitted a much more powerful instrument than physiological functionalism for studying multiple, interacting variables in internally regulated systems. The displacement of the organism from its privileged status within biology has been a triumph for neopositivism. The ability to study goal-direction, function, and signification entirely without necessary reference to living systems has removed a gnawing irrationality from the heart of organicist biology. Integration and adaptation have remained central, but they have become more thoroughly logical communications problems—on technical and ideological levels:

> Animal and Machine, each system then becomes a model for the other. The machine can be described in terms of anatomy and physiology. It has executive organs activated by a source of energy. . . . It contains automatic control centers to estimate its performances. . . . At any time, the machine that executes its program is capable of directing its action, of correcting or even interrupting it, in accordance with the message received.
>
> And vice versa, an animal can be described in terms of a machine. Organs, cells and molecules are thus united by a communication network. They exchange signals and messages in the form of specific interactions between constituents.[151]

It is time to rejoin primatology in the mid-1950s as a branch of technological functionalism, with its theories, practices, and institutional developments. Certainly not all of primate studies can be explained from this point of view. But the remainder of this paper will consider the thread in the primate story which theorizes nature as a cybernetic command-control system, i.e., a problem of communication.

Focus will remain on work connected to Carpenter's impact on primatology. That emphasis distorts the complex development of a broad field in which the relevance of individuals has become progressively problematic. Like Rocket-Man in Thomas Pynchon's *Gravity Rainbow,* the human components seem to disassemble and reassemble by a technological logic in the postwar primate project. But continuing attention to circuits linked to Carpenter allows initial exploration of themes that will need more refinement in an expanded context. These themes are: control of populations as a systems problem, social behavior as a communication system, excess aggression as a sign of stressed and outmoded systems with design flaws, and adaptation as a problem warranting educational and psychiatric intervention.

### *Stuart Altmann and the Boundaries of Communication*

Post–World War II primatology in the United States may be dated from the arrival of Stuart Allen Altmann, a Harvard graduate student, on Cayo Santiago in 1956. The coincidence with the National Research Council conferences on supply of primates, which culminated in founding the Regional Primate Research Centers, marked the two faces of the coin: institutional and theoretical development. Altmann, whose presence on Cayo Santiago was heartily supported by Carpenter, became the major student of primate society as a communication system based on stochastic processes. He saw all social behavior as communication behavior, in which behavioral patterns serve as social messages which regulate the behavior of other members of the group. He defined society as "an aggregation of socially intercommunicating conspecific individuals that is bounded by frontiers of far less frequent communication."[152]

Altmann was one of the first biologists to use the term *sociobiology* to refer to this point of view.[153] Finally, Altmann had his opportunity to study primates within the context of psychiatric interest in human adaptation, particularly in relation to problems of aggression. Conceptualization of the communication design properties of primate society set the framework for the study of design flaws, stress, and pathological system breakdown within a framework of cybernetic functionalism. These approaches were regarded with deep sympathy by Carpenter, and they represent a major strand in postwar primate studies.

In 1955 Altmann conducted a census of the howler monkeys of Barro Colorado Island. He adopted Carpenter's field techniques and regarded a natural population as an appropriate object in the study of control systems.[154] Altmann's Barro Colorado work was being done while he was in the Department of Medical Zoology in the Army Medical Service, at Walter Reed Hospital. David Rioch, the civilian director of the Neuropsychiatry Division with a special interest in problems of aggression and the neuroanatomy and physiology related to psychiatric concepts, facilitated Altmann's work.[155] In the general context of renewed interest in rhesus breeding colonies, Rioch had been looking for over 5 years for someone to be a scientific overseer for the ill-used Cayo Santiago colony.[156] Carpenter had enthusiastically contributed his 1930s census data on the howlers of Barro Colorado to Altmann. With Rioch, he was very interested in helping Altmann to study the rhesus of Cayo Santiago.[157] Rioch was clear about the motive for his interest: "My interest in this island was stimulated by the fact that . . . experiment studies in comparative social psychology would be of considerable value in view of increasing interest in clinical social psychiatry and preventive psychiatry."[158] Psychiatrist David Hamburg in the same years made similar points in beginning his collaboration with Washburn on emotions as adaptational complexes in evolution. Such complexes were subject to stress and breakdown in modern civilization.[159] Aggression and dominance were of special interest in this context.

Altmann sent his research protocol for the Cayo Santiago rhesus project to Carpenter in March 1956. He expected to spend 18 months to 2 years (1) determining population size, sex ratios, age distribution, disease problems, and individual identification; (2) conducting an observational study of behavior and social structure to obtain baseline data; and (3) performing intensive observational and experimental studies on facets of behavior, population ecology, and social relations. The first priority after studying population structure was an investigation of communication.

Altmann's organizing questions were an appropriate extension of Carpenter's semiotic-sociometric analysis to a framework of cybernetic functionalism. These were the questions underlying the budding science of sociobiology:

> What are the roles of the various sensory modalities in communication? What is the function of the communicative signals in the integration of the society? For every signal: what are the necessary, sufficient and contributory stimuli; what members of the society respond; and what is their response? What is the relation between communicative feedback and social homeostasis? Are there any social communicative networks that are "self-damping"? Does metacommunication exist? Are there (a) signals whose only function would be the "acknowledgement" of a signal emitted by another, (b) signals "asking" for a signal to be repeated, or (c) signals "indicating" a failure to receive a signal?

The section on "Reproduction" began with the question: "What are the behavioral signs of sexual receptivity?" That on "Social Dominance" was headed: "What is the dominant-subordinate structure of the society?" And under gregariousness, Altmann asked: "To what extent are the spatial relations between pairs of individuals an index of the amount of behavioral interaction taking place? Can changes in spatial relations be used to predict changes in social structure?"[160] In short, Altmann planned his questions so that reproduction, social structure, and communication mutually implied each other.

From his first rough draft as a Harvard student to organizing the international symposium on Communication and Social Interactions in Primates held in Montreal at the American Association for the Advancement of Science meetings in 1964, Altmann did the formative research and promoted the theoretical point of view that established the study of primate populations and societies as a problem in communication sciences. Altmann's work on primates provided the main material on the primates used by E. O. Wilson in his synthetic treatise *Sociobiology*.[161] The 1964 symposium, like the 1962–63 Primate Year at the Stanford Center for Advanced Study in the Behavioral Sciences organized by Washburn and Hamburg, marked the mid-1960s explosion in primate publications. The research base for those publications had been built from the mid-1950s. The two volumes resulting from the conferences also marked divergent points of view of the Carpenter-Altmann communications approach and the Washburn-Hamburg physical anthropology framework for promoting primate studies. The controversies go beyond this essay. But it is important that both conferences and volumes shared a basic interest in the problems of stress and their related technological-functionalist explanatory framework.[162]

Present at the Montreal conference were anthropologists (including the important Japanese primatologists), psychologists (including the major student of stress, Irwin Bernstein), psychiatrists and neurologists, a mathematician, a linguist, and zoologists. The interdisciplinary nature of primatology could not be missed. The anthropologist-linguist Sebeok followed Altmann's important theoretical paper on "The Structure of Social Communication." Sebeok defined zoosemiotics, the synthetic interpretive framework for the conference, in explicit terms linking the philosophic inheritance of the Chicago neopositivists, represented by Morris, with the communications engineering and information science assumptions. Study of primate systems was an investigation of communication with the help of a "typology of the sign systems used by the animals in the different sensory modalities at their disposal."[163] The relative informational and energetic properties of chemical, optical, tactile, and acoustic systems remained to be worked out and provided the research program for the future.[164] Sebeok called this work a study of language as a tool; zoosemiotics was literally a study of the technical design features of a system of communication. Altmann clearly described the tool as a structure of *events,* not a material *object.* A system of functioning signs was a logical behavioral structure, a patterned process, not a fixed object. Further, the study of communication behavior was the study of a control system that operated by stochastic processes. He considered that communication occurred when probabilities of the distribution of behavior were altered. They key concepts were energy and information and regulation of boundaries of communication. The translation of the physiological approach to behavior, analyzed in section 2 of this essay, into the language of cybernetic functionalism is striking.

Communication was an international concern. The Japanese figured prominently in this aspect of postwar primatology, and Altmann was a conduit for their work to an English-language audience.[165] Carpenter also was in contact with the Japanese research program in primatology from the early 1950s. The Japanese research drew on his prewar field studies, as well as on the work of Robert Yerkes and H. W. Nissen, but developed quite independently at the Japan Monkey Center, established in 1948. Researchers at Kyoto University began intensive observations of the Japanese macaque in 1948 and formed the Primate Research Group in 1951.[166] In 1966, after the Eleventh Pacific Science Congress held in Tokyo, Carpenter had an opportunity to study the Japanese macaques.[167] In 1969 he attended a conference at the University of Missouri held to consider sites for a transplant of a Japanese troop to the U.S. Primate biologists at the University of Wisconsin (John Emlen), Rockefeller University (Peter Marler), Osaka City University, and Kyoto University had initiated a cooperative research program on communication mechanisms in *Macaca fuscata.* The complete troop of 100 known individuals was being offered to facilitate continuing studies in primate sociobiology. Again, sociobiology was clearly tied to the development of research into animal groups as communication systems. Let us pick up this thread in a grant proposal by Emlen in 1971 in order to summarize both the interconnections of research personnel in this approach to primatology and the theoretical and practical constitution of primate society as a command-control system.[168] Carpenter was a referee for the grant proposal to the National Science Foundation, "Group-Specific Communication Patterns in Primates," submitted by Emlen and Gordon R. Stephenson.

### *John Emlen and the Sociobiology of Communication*

Emlen was an ornithologist trained by Arthur Allen at Cornell, earning his doctorate in 1934. At the University of Wisconsin since 1946, he was critical in the development of primate studies as a modern evolutionary science and its inclusion in the sociobiological synthesis. His graduate student George Schaller did the first extensive field study of the mountain gorilla in the early 1960s.[169] Emlen was deeply involved with Harold Coolidge, Washburn, Carpenter, and others in establishing primate studies in Africa, beginning about 1958. In 1960 he wrote a comprehensive plan for African field studies that combined ethology, ecology, general life history, and social organization.[170]

The 1971 grant proposal with Stephenson was a full summary of the sociobiological-communications point of view in primate studies. Stephenson and Emlen proposed a 3-year investigation of *M. fuscata,* cooperating with the Japanese in collecting data from three troops in Japan and developing the most sophisticated techniques for data analysis at Wisconsin. In basic theory and in concrete research design, Emlen and Stephenson combined physical communications engineering and the theory of signs. The ultimate problem to be explained was the systems "emergent," i.e., nongenetically transmitted culture.

> It is now generally accepted that a set or system of communication patterns constitutes the primary integrating mechanism in complex societies and that complex social environments, in turn, provide the basic conditions for the emergence and development of culturally or socially (vs. genetically) transmitted behavior. Recent studies have demonstrated that several nonhuman primates have the capacity to acquire communicatory behavior patterns in this way. . . . Anchoring on our current analysis of signal form, the contemplated research will examine the effect of syntax on communication patterns in their functioning as mechanisms of integration among individuals in the complex societies of *[M. fuscata],* the only nonhuman primate for which cultural differentiation has been described in detail. . . . A fuller understanding of these phenomena would have valuable implications for problems of human sociobiology.[171]

The proposal began with a consideration of semiotics inherited from Peirce and Morris. The study drew heavily from Altmann's previous work on rhesus.[172] The syntactic dimension of semiotics (formal relation of sign vehicles to one another) was defined for this study as the sequence of signals or intercalation of signals from more than one communication modality. The question was: Do particular signals occur in syntactical relations as a function of context? This kind of analysis made constant use of the Shannon-Wiener theory of information, which applied in particular to the syntactic dimension of the theory of communication. The monkey interaction system chosen for this study was, appropriately, the male-female consort pair. The immediate reason was that different observers could agree with a high degree of reliability in recording sexual interactions. An implicit context was the continuing importance of the phenomena of reproduction in explaining social groups within the framework of sociobiology. The data for the study "were vocalizations and stylized non-vocal behavior monitored by human observers as they occurred in the course of social interactions among Japanese macaques. These data are treated as communication events which occurred in a communication channel maintained among the individual macaques."[173] The basic data concerned mutual molding of behavior through development of systems of signs.

Stephenson and Emlen summarized the purpose of their proposal as an effort "to establish much more precise relations between signals and contexts in order to elucidate the role of syntax in the communication system of Japanese macaques."[174] The research would produce a refined catalogue of behavioral pattern types that partitioned communication events "in the same way as do the monkeys" and a "typological frame of reference for recording statistically sufficient numbers of communication events in natural sequences in order to develop a set of rules of usage which describe the basic communication system of Japanese macaques." The final objective was "to examine the role of Shannon-Wiener information on the syntactic dimension of semiosis in the specification of expectation on the semantic dimension of semiosis among Japanese macaques."[175] The language in the proposal itself presented a major communication problem!

The hardware of the proposal matched the theory. The Wisconsin group had developed a SSR system recorder,

> A keyboard event recording system with computerized transcription and analysis of data entries. . . . (W)e have conceived, designed, and finally built and tested an event recording system which meets the following criteria: a recording system for field use . . . ; maintenance of the integrity of data in terms of real time, subject (*S*ender), action (*S*ignal or behavior pattern), and object (*R*eceiver), and the incidence, duration, coincidence and sequence of data entries; high flexibility, hence key assignment through software rather than wiring; ready detection of errors; automatic transcription of categorical data; low cost, and simple operation.[176]

Compared to Carpenter's methods of making field notes, data recording within cybernetic functionalism was in a different world. The intuitive scatter diagrams and sociograms gave way to the computer programs that were several orders of magnitude more powerful as tools in studying systems of self-regulation. The similarity in goals has been masked by the disparity in technology. Functionalism persisted, but physiology gave place to a more powerful instrument of control.

Before the war Carpenter had adopted semiotics within a framework of physiological functionalism to explain and facilitate social integration. The tie between that research and the Emlen-Stephenson proposal was direct. Transformed by the apparatus of cybernetic functionalism, the problem remained the study of animal societies and populations as primordia, anlagen, or emergent communication systems in reference to human culture and promotion of human adaptation. Just as the relation of biological and human sciences was an important dimension in the analysis of the Carpenter work, that issue has remained central in the postwar development of sociobiology. A conclusion of this essay is that since 1930 the direction of "borrowing," "reduction," or "bridging" is *from* the human sciences *to* the biological sciences. Primatology, as part of the sociobiological synthesis, must be seen primarily in terms of extension of the systems sciences developed in philosophy, linguistics, labor management (ergonomics), and military communications engineering to an explanation of nonhuman groups. Most public attention has been directed to sociobiology's claim to explain culture (especially human kinship and cooperation) in terms of the expanded theory of natural selection (inclusive fitness). Genetic "reductionism" thus has been the focus of complaint by social scientists.[177] But communications "reductionism" might well be much more basic to sociobiologists' claims—and a much more serious challenge to the Boasian culture concept, in the form, as Wilson noted, of cannibalism.[178]

From about 1910 the polemics for the Boasian culture concept against the remnants of American social Darwinism attacked the reduction of culture to physiology or its extensions (superorganism).[179] But the functionalist logic of Boasian social science proved entirely compatible with a reformed biology that accepted the interactionist paradigm (heredity-environment) without quibble. That theme was explored in detail in section 3 of this essay. The functional-structural social system school that grew from Henderson (e.g., Parsons), the social psychology represented by sociometry, and the functionalist linguistic philosophy of semiotics were all part of the rejection of biological reductionism. Both the biological and the social sciences operated by an organicist-functionalist logic of precybernetic systems. The post–World War II redefinition of culture as a communication system, and therefore a problem in design analysis of a system of signs, has been a consistent development of the prewar interactionist paradigm.[180] For Emlen and Stephenson, representatives of the sociobiological development of the synthetic theory of evolution that was an essential part of the peaceful relations of the biological and social sciences by 1940, culture was not some kind of idealist structure over and above individuals. Rather culture was an emergent of communication patterns between and among individuals. Human patterns, and those of some nonhuman primates, were characterized by a large degree of openness, or arbitrariness. The evolution of such open communication systems has been a major theme for the spokespeople of the synthetic theory of evolution.[181]

Emlen and Stephenson expressed the issue clearly:

> With the emergence of culture as a primary adaptive mechanism in some species of primates, maintenance of the social group *per se* gains further biological significance. . . . This has direct import for human sociobiology where it is now recognized that for humans the individuals of the social group maintain and perpetuate cultural behavior via communication between and among their fellows (in contradistinction to the notion of the "superorganic". . . . The capacity for socially mediated acquisition of arbitrary signs has been regarded as *the* prerequisite to the emergence of culture in the hominid line. . . . The writers of this proposal have been concerned with the implications of these data for the basic concept of culture and the nature of processes mediating social and ecological integration among human societies. We have accordingly organized and pursued a program of study into the general sociobiological mechanisms of cultural diffusion and maintenance in primate social groups.[182]

The irony of social scientists taking refuge from the claims of sociobiology in a culture concept based on language is immense. The human sciences have been systematically constituted as an organon of social control. That the tool can destroy its maker has been an old theme in the primate story.

### *Carpenter's Practice of Communication*

Carpenter was not the most important figure in postwar primatology; indeed there could be no such individual under the conditions of large-scale "industrial" biological research to which Weiss alluded in 1955 at the National Research Council conference on primate supply. But Carpenter still provides a useful strand into the web of primate developments. It is appropriate to conclude this essay by returning to him. He was left in the midst of a grant application in 1947 for a teaching aid device made possible by wartime communications research. Let us bring him from the discouraging years for primate students, through his own development in the 1950s and 1960s, to his last associations at the University of Georgia and the Yerkes Primate Research Center in Atlanta.[183]

Throughout the 1950s and 1960s Carpenter wrote synthetic articles based primarily on his prewar fieldwork and on some additional fieldwork on Barro Colorado howlers in 1959. Emphasis continued on the familiar categories of socionomic sex ratio, coordination, integration, territory, etc. But the language of the papers depended more and more on cybernetic theories, techniques, and metaphors. He placed more stress on seeing primate groups structured by stochastic processes, and his statistical methods became more sophisticated.[184] He believed primatology properly studied the establishment and maintenance of "steady states" and of "sociostasis" analogous to homeostasis.[185] He analyzed the familiar dominance or status hierarchies in terms of the degrees of freedom possessed by individuals as a function of their position in the system. The entire system functioned to reduce individual and social stress.[186] Carpenter understood the entire question of group coordination in terms of the question: "What kinds of behavior serve the functions of communications as processes of control?"[187] Statuses were perceived as signals in a communication system. Design limitations in the nonhuman primate communication systems accounted for the differences between human and simian societies.

Carpenter attempted to conduct new observational and experimental studies on the design features of rhesus communication systems. Together with the Cornell linguist Charles Hockett, from whom Altmann drew heavily, Carpenter planned work on Cayo Santiago in 1965–66. Altmann believed that the "universal design features" of human language catalogued by Hockett were all present in nonhuman primate communication systems. The design features were drawn from cybernetic communications engineering: channel of communications, broadcast transmission, directional reception, directed messages, multiple coding potential, total self feedback, semanticity, arbitrary denotations, and others.[188] The laboratory director refused Carpenter and Hockett access to the island colony on the basis of their failure to complete application forms in time. Clearly, Carpenter was not an independently powerful figure in postwar primatology; he was one of the many primate students caught by the excitement of the international rebirth of the field from the mid-1950s.[189] From the 1930s Carpenter had viewed populations as self-regulating systems. Here again, his language became less physiological and more cybernetic, or part of "general systems theory."[190] Maintenance of the species-specific norms for grouping patterns was important in controlling "stress." "Furthermore, deviations from these typical groups are correlated with motivation and behavioral mechanisms ('stress') which operate to reestablish the group norms."[191] The concept of stress tended to replace the prewar concept of frustration. This seemingly innocent change in language signified an important metaphoric transition from physiological hydraulic models, where damming of organic energies could produce leaks and burst pipes, to other physical systems models more compatible with a cybernetic analysis.[192]

Territoriality was a behavioral regulatory mechanism critical to maintenance of optimal population states. Carpenter explored that theme in his paper in the 1958 Roe and Simpson volume, *Behavior and Evolution,* which extended the neo-Darwinian synthetic theory of evolution to comparative psychology. "Territoriality combined with social organization reduces stress, conflict, pugnacity, and nonadaptive energy expenditures."[193] Carpenter picked up the theme repeatedly, and there was a ready audience. The concept of design features of primate aggression mechanisms in relation to large-scale system breakdown got particular attention in the United States during the Indochina war. Carpenter contributed a paper on "Territoriality and Aggression in Nonhuman Primates" for the 1967 conference at the Pacem in Terris Institute at Manhattan College. His lesson was that territory was a conflict-reducing regulatory mechanism in nonhuman primate communication systems. The problem was one in systems management.

Carpenter was even more explicit on the cybernetic model for understanding aggression in his contribution to the Symposium on Anthropology and War held in Washington, D.C., in November 1967.[194] This was one of the few occasions where Carpenter argued that a nonhuman primate characteristic was *homologous* with human behavior. His main proposal was "that we describe the *design characteristics* of aggressive behavior in a manner similar to that of scholars in linguistics who describe the design characteristics both of languages and of primate vocalizations."[195] He argued that aggressive behavior functioned to regulate grouping structures and was an essential part of the regulatory system of animal groups. Without aggressive behavior, there could be no territories, dominance hierarchies, group structure, or inclusion or exclusion of young. Stimulus-use-sign networks of communication could not operate without aggressive behavior, but signaling could replace physical aggression. Excessive aggression was maladaptive. "What behavior is adaptive or maladaptive must be judged by criteria from the biotic levels of ecosystems and defined population dynamics."[196]

In terms of animal population management, control of the number of males in groups could provide desired control of maximum reproductive potential. In nature, increased male fighting could accomplish the same end; i.e., "the efficiency of population biomass has improved reproductive efficiency."[197] The problem before the symposium was how "man" could use "his" brain to "design strategies for regulating aggressive behavior." Carpenter emphatically meant *regulate,* not *eliminate,* such behavior. He searched the animal world for helpful regulatory models, noting that the homology in aggressive behavior between animal and human occurred at the level of endocrinological-neurological mechanisms. Attention to design characteristics at this level was necessary. It seems Carpenter was not being simply metaphorical when he called for people to use their brains in regulating aggression. This matter will be considered further in the conclusion to this section.

Usually Carpenter did not argue that human systems were identical to or homologous with nonhuman ones. Rather, primatology studied "the evolution of man and his analogous behavioral and ecological adaptations."[198] That simple statement was made in the introduction to a volume of papers from the Eighth International Congress of Anthropological and Ethnological Studies, held in Tokyo in 1968, and the Second International Congress of Primatology, held in Atlanta in 1968, where Carpenter moved a year later. Carpenter organized the volume entirely around the theme of cybernetic processes, including in the theoretical section contributions by the former student of Emlen, Stephenson ("Biology of Communication and Population Structure"), and by the man with whom he had done the 1959 howler census, William Mason, then of the Wisconsin Regional Primate Research laboratory ("Regulatory Functions of Arousal in Primate Psychosocial Development"). This book represented the culmination of Carpenter's publications on primates. The concluding statement of his introduction expressed his full transition to cybernetic functionalism in conceiving the relation of primatology to human sciences:

> Research on the complex social behavior of nonhuman primates is aimed at the deepest understanding of these primates as evolutionary products and as progenitors or parallelisms in man. These studies of *complex systems of behavior* include both the regulators and the regulated behavior . . . [and invite] the application of *behavioral systems analysis* and challenge investigators to develop improved methods for accomplishing these complex studies. A central feature of most biological systems, including behavioral systems, is the cybernetic or "feedback" processes. These reciprocal communicatory processes operate as adaptive regulators just as surely on the complex ecosocial plane as they do on the physiological levels of nonhuman primate life.[199]

In working for the integration of human primates, Carpenter followed the logic of control he had found in analogous systems. For over 20 years he worked in research, development, and application of educational technologies in India, Japan, Guam, Spain, and "developing regions of the United States."[200] He looked for the "minimum required compromise" between the logical blueprint for engineering adaptive instructional technologies and the "irrational" impediments of sociopolitical realities.[201] Media were "applied" in an engineering task of producing adaptive educational innovations for planned development. Once a "target country" was selected, a rational study should produce the proper recommendations to match the learning need to the technology. Demonstration or "proving ground mechanisms" should be established. "Learner populations" would be tested for their characteristics. "Indeed, the task is so formidable that actual test-and-fit operations are parts of modern advanced methodologies for producing effective instructional programs." He perceived the target population as a moving population, so that changes in attitudes (always the kind of learning he considered most seriously) should be tracked throughout. Evaluating the "communicative power" of a particular modality was very difficult. "Likewise, this task is especially complicated when basic and deeply embedded attitudes and cultural norms are targets of behavioral change effects."[202] He was referring to attitudes toward childbearing and childrearing and family life—the area most critical to social control for Carpenter. It was all a problem of operations research—ideal gunnery control.

Carpenter emphasized the need to avoid biases, by which he meant conceding, where not absolutely necessary, to sociopolitical pressure. His inheritance from Chisholm was firm. The task of the communications engineer was one of "convincing decision-makers to accept and support engineered and rationally designed plans."[203] The communications consultant was a specialist in attitude engineering. Carpenter worried that such engineers became too committed to sweeping high technology strategies, and did not pay full attention to the multimedia systems concept, with many kinds of communication channels. They should be evaluated in terms of cost-efficiency energetics. He also worried about "interference" from "information overload," leading to his concept of "literacy" for developing countries:

> This definition includes receptivity to many different channels through which may flow information for learning. . . . This problem of defining characteristics of *prevailing learning sets of populations* is worthy of new research. In brief, what may be termed cultural patterns of learning styles need study in order to relate them appropriately to small technologies applied in developing countries to modify and improve the adaptiveness of peoples' behavior.[204]

The cybernetic, sociobiological logic of control based on communication was explicit. Producing literacy did not mean nurturing such basic skills as independent reading; it meant use of the channel or communication modality that permitted most economical control of learner attitudes in the interests of adaptation defined and engineered by an outside specialist.[205] Indeed, evaluation forms for determining "quality" of an instructional technology considered only "those factors which produce the desired behavioral changes in the target population." Within a logic of cybernetic functionalism, for both human and nonhuman primates, sociobiology was developed as a synthetic communications science of integrated control systems.

### *Conclusion: Carpenter's Last Words*

Carpenter's knowledge of primates began with the wild howlers of Barro Colorado Island under the tutelege of Robert Yerkes, a man whose life was dedicated to human engineering through the production of primate models, and Chapman, a conservationist and ornithologist. His last field research, in 1971, was on brain-implanted, telemetry-controlled gibbons on another island research station, the Bermuda Primate Center, on Hall's Island. This time his fieldwork was made possible by the human psychiatrist and specialist in technological control of aggression Delgado, who like Yerkes, was based in physiology at the Yale University Medical School. And here Carpenter operated under the intellectual influence of another ornithologist, Emlen, a major builder of cybernetic-communications sociobiology in bird and primate work. As in 1930, he wanted to understand social bonds, especially sexual bonds, in order to promote cooperation—in order to facilitate communication. It is fitting to close this paper by following Carpenter to Hall's Island in the summer of 1971.

Hall's Island was a 1½-acre piece of land near the Bermuda Biological Station, a marine and oceanographic research laboratory established in 1910. The island was rented from private owners at the end of 1969 for the Bermuda Primate Center. The center was actually the arrangement for gibbon research by Delgado, of Yale; A. H. Esser, the director of the Social Biology Laboratory of the New York Rockland State Hospital; and N. S. Kline, director of research at Rockland. Delgado, Esser, and Kline were also associated with the Psychiatric Research Foundation in New York, an organization founded to promote controversial investigations in psychopharmacology. The Bermuda Biological Station approved Esser and Delgado's gibbon research plans and cooperated by providing support facilities. The Public Health Service provided about $130,000 for a 1970–72 investigation of "Telemetric Control of Free-Ranging Gibbons."[206] Carpenter, principal investigator on a related proposal, observed the 5 juvenile and 1 adult gibbons on the island from mid-June to September 1971 and made a series of recommendations for the improvement of behavioral study.[207]

Esser's grant proposal clarified the context of the research:

> Life in a group provides protection for the individual. However, the group member has to pay a price for this protection; he has to submit to the social order. In today's society there are increasing numbers of people unwilling or unable to pay this price; for such people the guidance and restraint traditionally supplied by family and classroom is no longer effective—we call these people alienated. Social planning cannot hope to reduce alienation if it unwittingly flaunts the deep-rooted biological laws which underly social cohesion. Utopias are notoriously short-lived; in our troubled cities today the traditional forces of governmental action give little relief, yet radical proposals are justifiably suspect. We lack the necessary knowledge of man as a social animal to evaluate our social plans in advance. . . .[208]

Carpenter's belief that human and animal aggression were homologous at the level of brain and endocrine mechanisms justified his part in this research plan. In the 1960s he and Delgado had both participated in the plethora of conferences to consider war, aggression, stress, and territoriality. But further, gibbons had a social structure that seemed especially interesting for human beings at the time of the Asiatic Primate Expedition in 1937: they defended territories and lived in monogamous family groups. The Delgado-Esser application did not ignore this issue. They wished to study naturalistic populations, more relevant to modern human stress and alienation than the prison and mental hospital patients and laboratory monkeys to which they had access ordinarily. Recognizing constraints on human experimentation, they sought as an experimental system a free-ranging primate with important social analogies to "man": "It appears heuristically fruitful to investigate spontaneous social behaviors in that species which shows the greatest similarity to what appears to be basic human social structure. For this reason we decided to use the gibbon *(Hylobates lar),* the only ape showing territorial behavior while living in a nuclear family constellation."[209]

The proposed research was a straightforward extension of work Delgado had done for over 20 years. He had been instrumental in developing the multichannel radiostimulator, the programmed stimulator, the stimoceiver (air force funding, used for patients with psychomotor epilepsy), the transdermal brain stimulator, a mobility recorder, chemitrodes, external dialtrodes, and subcutaneous dialtrodes.[210] These were the organs of cybernetic functionalism. The Hall's Island study would provide the needed correction of laboratory data with field investigation. Behavioral observations would be automated as far as possible, using the SSR device developed by Stephenson while he was a graduate student under Emlen at Wisconsin. The point of the island research was to gather baseline behavioral data on operated and unoperated gibbons and then to stimulate selected brain areas electrically and chemically by telemetric control and observe the ensuing alterations in aggressive behavior. Such behavior was a basic part of the mechanisms maintaining territorial integrity in gibbons. Its spontaneous occurrence was ensured by the conditions on the 1½-acre island, since ordinary gibbon territories covered 30 to 400 acres.

> The main purpose of the project is to investigate the possibility to induce lasting modifications of free-ranging behavior by means of longterm stimulation of the brain. . . . We hope that our project will increase the knowledge of the cerebral bases of anti-social behavior, and thereby contribute to a better understanding of the many factors involved in the study of this important part of our social life. . . The methodology and technical expertise for the behavioral aspects of the project have evolved over the past nine years in continuous studies of the behavior and biochemical functions of a group of schizophrenic patients in the Behavioral Research ward of the Research Center at Rockland State Hospital.[211]

From his early connections with reproductive and neural physiology at Stanford, Carpenter had been consistently interested in correlating brain and social structure research. Compared with the earlier organicist physiology, the context of cybernetic functionalism expanded the technical power and the theoretical scope of communication as control. When researchers on Cayo Santiago had complained to Carpenter in the mid-1960s about extensive brain lesion work going on in such a way as to destroy the integrity of social groups needed for naturalistic observations, he responded, "I think that a reasonable amount of intervention operatively on the free-ranging primates has always been in our plans."[212] Carpenter agreed that Cayo Santiago needed an advisory board to help coordinate research priorities, but his weak defense of the behavioral biologists was a major disappointment to them.[213] On Hall's Island, Carpenter was upset by the situation, because he felt insufficient care had been exercised in keeping baseline behavioral data on social interactions before operations. He did not object to the basic research goals, but directed his recommendations to improving the research to meet them. He recommended the research program be extended to at least 5 years with 3 or 4 pairs of gibbons. Noting that Delgado and Esser wished to develop an "early warning system" for potential suicides and "depressive psychotics," Carpenter reported, "They hope, through the Hall's Island project and parallel experiments in laboratories in the States to get to the stage where a patient in a hospital can be given a battery of psychological tests—including a period of MTS (Mobile Telemetry System) and results would then be fed into a computer which would spell out the correct medication."[214]

Carpenter's last fieldwork was a fitting conclusion to a career that began with the study of monogamy in pigeons and socionomic sex ratios in wild primates. On Hall's Island, he was a consultant in the communications engineering of aggression, a regulatory behavioral mechanism in the desired steady-state population dynamics of human and nonhuman primates. Aggression was part of a sign-signal-control system whose design properties could be known—and then reengineered to prevent stress and maladaptation. Surely, many primatologists have not shared all of Carpenter's theoretical and practical commitments. These important differences require more examination. This paper merely claims that his work fits into a larger controlling logic which has undergone a transformation from physiological to cybernetic functionalism. That logic is pervasive throughout the biological and social sciences; sociobiology is only one normal elaboration of a common raw material. Carpenter's work would not have been possible without the ideological, technical, and institutional bases this essay has briefly explored. He was one of the individual founders of modern primate studies; he was also one of the engineered components in the constitution of a science of dominance as the organon of human life together. In Robert Yerkes's terms, Clarence Ray Carpenter was a servant of science.[215] In the beginning was the word.

## 5. Primate

For thus all things must begin, with an act of love.
—Eugene Marais, South African naturalist, 1980

In the introduction to this essay, nature was called fact and fiction, and in both dimensions of making, a cultural production. The first order in nature, that which modern Western natural scientists since Linnaeus have named primate, has been a contested cultural object since its beginnings. Primates are scientific objects that embody, echo, and represent the major dualisms and political struggles of the tortured relations of knowledge and practice in modern societies. Primates particularly illuminate struggles over the distinction between nature and culture.

People are primates; people named this fact. Primates are an order in nature; that is, primates are scientific objects of knowledge that have been historically produced as productive internally controlled systems, as functioning wholes, as reproducing bodies. Primates are biological objects, even in their recent technological metamorphosis as cybernetic command-control systems. Primates were also biological objects as physiological energy-processing systems. Distinctions between technology and biology have never been at the heart of belabored and beloved distinctions between nature and culture. Nor have distinctions between political, economic, and biological principles of order been at the heart of the transformation from animal to human. Fueling the motor of that transformation is another distinction altogether: between dominator and dominated. The nature/culture distinction is about logics and practices of domination. That is the cultural stake in the structure of scientific natural objects, among whom we number ourselves as primates. The primate order has been built to embody particular kinds of internally functioning wholes depending on nature as resource, as pilot plant for human engineering, and as regulated reproductive body. In contrast to other perspectives, the order seen and produced in biological practice depends on internal workings of power (organisms, systems), not on externally or forcibly imposed rule. The extraordinary legitimating power of biology for modern forms of domination is rooted in this fact. Biology produced natural objects ruled by principles of rational internal control.

Primatology has produced primates. This essay has attempted to demonstrate in detail the dynamic logics of particular biological objects in order to illuminate the underlying structure of the orthodox distinction between nature and culture. This essay has insisted that the structure and locus of that distinction has little or nothing to do with academic boundaries between life and human sciences, with charges and claims of reductionism, or with separation of "man" and "animal" into objects of anthropology and biology. From physiological to cybernetic systems, primates between about 1930 and 1970 emerged with an internal command structure fully supportive of the proper relations of nature and culture, serviceable to man. Nature/culture is the internal scaffolding of all the life and human sciences; the dualist principle of domination does not assign one pole to biology and the other to human social science.

Until very recently most Western writers, especially anthropologists and philosophers, have asserted that some form of the distinction between nature and culture is universal and essential to being and becoming human, to transforming the raw material of nature into the productions of culture. Nature/culture has been voiced as an ontological, not merely analytical, distinction. Under the pressure of developed oppositional voices from women, people of color, and all the others who did not previously name the major dualisms that were used to produce and interpret nature and their place in it, this fundamental dualism has become unstable. Nature, the other which must submit to the probes of knowledge and the practice of husbandry, is being deconstructed and revoiced.[216] It is perhaps now historically possible to craft a nature not structured by principles of dominance and practices of domination, to know something other than the natural order of command-control systems.

But this paper must end with the old order. The documentary filmmaker Frederick Wiseman took his camera to the Yerkes Regional Primate Research Center in Atlanta to provide a vision of a system of production of modern science. His cultural production, *Primate,* is a fitting conclusion to "Signs of Dominance."[217] The fit derives not only from the presence of major actors in this story at the primate institution (e.g., Altmann), the roots of the laboratory in Robert Yerkes's Orange Park facility, and the importance of the regional primate centers as a resolution of 1950s supply and research crises sketched in the introduction of this essay. More critically, Wiseman sees the laboratory as a functioning whole, as a cybernetic organism internally ordered by the logic of dominance working to give birth to fully controlled, rational, reproducing systems. Wiseman has ground a distorting mirror to tell an origin story. The scientists at the Yerkes center have deplored Wiseman's film, and my purpose is not to defend its accuracy or fairness. My purpose is to examine its own symbolic story, its vision of the production of nature as cultural artifact with a phallocratic birth. "For thus all things must begin, with an act of love."[218]

*Primate* has the structure of an origin story and should perhaps be read as science fiction, as well as social commentary on the production of scientific fact. At the least, *Primate* is a myth of modern self-birthing, of self-preoccupation, of the achievement of man's humanist goals of self-knowledge in science and technology. The film opens with a hall of portraits of human bearded male fathers of primatology, from Linnaeus to the present. As we shall see, the film closes with another sort of male-produced birth, in which animal, technology, and human being are co-engineered to gestate in space. In between, Wiseman envisions a system of production of scientific knowledge in which animal, machine, and human are integrated in a self-regulated techno-organic whole. The dominance hierarchies of species, sex, class, and race are all pictured as components of a system whose product is *Primate*. The film begins with animal birth and ends with technological birth, and the two stories are one reproductive whole.

After the portrait gallery, the camera takes the viewer up the drive into a complex of scientific buildings, to a hall of cages, where filmmaker (bearded) and scientist (bearded) discuss gorilla sex in front of a male-female pair of that ape species. The sound track gives us a fragment of discussion on data collection in ape sex research; in particular, the scientist explains the observation and caging system: "We don't want them doing things when we can't see it." In the office of the scientist we see photographs of copulating gorillas and get a discussion of sexual behavior observed in the field by George Schaller. The theme of sexual control has been unmistakably announced.

But the film will not explain these matters externally; we must learn to see the whole from within. *Primate* is a film about the structure of observation, about how to produce knowledge, about daily practice in a scientific laboratory, about objectivity, about alienation. Each scene dissects the means of production of what will be allowed to be seen. A major theme is the distancing of observations, the structuring of vision. *Primate* has been criticized as a film about scientific sadism; that judgment is radically wrong, in my opinion. It is a film about the production of objective vision, about mind over body, culture over nature, the whole over the sum of its parts. There are no cruel human beings in *Primate;* indeed, Wiseman edits his film to evoke compassion and anger. From the opening sequence of photographs and portraits, observers in front of cages with tape recorders and standardized data sheets, and pictures of gorillas discussed in front of a visible filmmaker, to the ending scenes of a technological birth watched on radar, there is no immediate vision in *Primate*. There can be no illusion of immediate nature in science.[219]

Next, the camera takes us to a birth scene; a newborn orangutan crawls clumsily over its mother's bloody head. She sits placidly sucking the placenta and stroking her infant. Outside the cage a white-coated white female technician takes notes in a ruled fashion while a white male scientist speaks into a tape recorder, translating the physicality of the birth into frequencies of motion in space. We are then taken into the infant reception room of the laboratory. A black woman feeds a baby ape with a bottle; a white woman rocks an ape baby in a chair and croons "mama's baby." But this is no scene of unleashed maternal nature. All the responses of apes and humans are integrated into a data collection system, where hugs and kisses are ways of relaxing infants for insertion of thermometers and translations into marks on paper. At the end of the scene a white male human enters to check charts and rapidly scan the room; he does not interact with other humans or animals.

The next sequence shows young apes returned to their cages after a period in the exercise yard. Several of them cling onto the body of a black man. There is no interaction.

The saga of birth and growth appears to be interrupted in the next sequences; but the fragments will come together. A rhesus monkey has a blood sample taken; an automated blood analysis system is seen; a black woman does hemoglobin counts in a microscope. Two white men discuss the use of a stimulus device to get semen samples from the chimpanzee John. At one frequency, the technician generates an erection, at another, an ejaculation.

Shortly after we see a male chimpanzee taken from his cage, anaesthetized, and stimulated. Five men hover over to assess the amount of semen, while another cleans the teeth of the insensate animal.

Many fascinating scenes cannot be discussed here; Wiseman provides a multiplicity of visions, taking us into the language experiments with the chimpanzee Lana; into a trailer to observe locomotion in apes to study evolution of bipedalism; into cerebral localization laboratories; and into chambers for studying the physiological effects of weightlessness. Throughout, we walk down hallways where black women empty garbage, white women receive orders from white males on preparation of samples. The laboratory is a workplace with a social division of labor. We do not know the product of this workplace yet; we see originally only fragments. No one tells us the whole truth. But purposefulness is everywhere.

Three major sets of sequences must be discussed. They do not come at once. They originally appear in bits, but assemble into a system of meanings, a myth. The three sequences concern experiments on brain localization and sexual and aggressive behavior, on gravity effects, and on artificial insemination. The first two studies use rhesus monkeys, the last, chimpanzees. All use human beings and elaborate technology.

The experiments on cerebral localization and behavior bring us back to the familiar territory of Delgado and Carpenter on Hall's Island. There is nothing sinister in these experiments; they are the daily study of brain function and its internal control. *Primate* shows the viewer two white women using a banana reward to insert electrodes into a box installed on the head of a rhesus. The women must be new at their task; they and the monkey emit frequent fear grimaces in repeated unsuccessful attempts to connect the hardware. Finally the preparations are complete. The male rhesus with box is released into a test cage, and two females (reagents?)

are added. Scientists in another room full of electrical equipment and recording devices turn the monkey on and off to elicit and damp out aggressive behavior. The visual attacks and pauses are translated into data of unit behavior frequencies and stimulus strengths. This sequence set is perhaps one of the most eloquent pictures in the film of a communication system animated by a teleology of control. Human, animal, and machine work smoothly in mutual communication. On, off. The laboratory is a pilot plant for human engineering. Nature is a design engineer. The goal is communication control.

Only gradually could a viewer conclude that another sequence set concerns gravity variations and their physiological effects—and even more gradually this work emerges as part of the U.S. space program. The viewer sees a room crammed with equipment—viewing screens, automatic recorders, dials. Nothing is explained; all the other inhabitants of the room seem to the viewer to know what is going on—except, we see, one other. A lay white woman in street clothes, perhaps a reporter of some kind, seems to be visiting. Her squeamish, mystified face witnesses the insertion of a rhesus strapped in a standard lab restraining chair into a large black box. The woman seems in the way; she dodges, watches uncomprehendingly. She asks dumb questions; and a white male scientist talks about baseline data before the flight, about mimicking lunar or Mars gravity. The door of the black box is closed; the monkey spins faster and faster inside. Human beings watch the monkey intently—on a television screen which shows the animal's contorted face. Various recording channels have been activated; data are produced from several inscription devices in long streams. The multiple layers of observation, the scientific metaphor of a black box, the relations of lay woman and male scientist, the monitored womb for generating data: these are the components of Wiseman's assembling primate system.

It is time to return to the chimpanzee John and the careful efforts to get a semen sample for artificial insemination experiments. Wiseman leads us into another kind of scientific space, the seminar-conference room, for one of the funniest glimpses in recorded history of scientific intercourse. The room, devoid of machinery, plainly furnished with a long table, is full of white men of all ages conversing about schedules for obtaining John's samples for inseminating Flora, Cherry, and Banana. The viewer hears they have frozen sperm for a backup, hears cost-benefit analyses of sampling on Tuesday or Wednesday, hears all the details of experimental logistics—without, of course, ever hearing anyone tell the invisible viewer why these preparations make sense. Wiseman reserves that conference for later in one of the meanest sequences in the film. This first scientific conference ends with the overheard phrase summing up the results of the deliberations, "Let nature take its course." An acne-faced young man exits the room. He will reemerge some sequences later, a lab bottle full of grape juice in one hand, a tubular stimulus device in another. He will entice John with grape juice, giving him squirts while conducting a scientific masturbation with great skill.

But what is to be born in this film? Is it perhaps simply sophisticated scientific S and M? Highly objective pornography? Or is this film a vision of a major system of production in modern society? To what end is the power of this film? Why this elaborate labor? These questions demand that we proceed to the last two sequences of the film, to at least one reading of the payoff from this laboratory. The laboratory is not sterile; it produces.

That, anyway, is the lesson stressed by the director of the laboratory, Geoffrey Bourne, in the penultimate scene of the film, the second scientific conference Wiseman provides the scientists for explaining their rites. The inheritor of Robert Yerkes's scientific legacy,[220] Bourne appears a kind of elder or priest sermonizing the workers in the lab in a period of funding crisis to motivate them to fight for the material foundation of their efforts. Bourne in 1974 and Weiss in 1955 are members of the same fraternity nurturing primatology. Bourne addresses an all-white male audience on the relations of pure and applied research, the frontiers of research, and the threat of displacement of America's lead in biomedical science by European competitors. He calls on his ancestors, Abraham Flexner and Sir Alexander Fleming, a man of vision who (with Rockefeller money) modernized the structure of medical-scientific research and a winner of a Nobel Prize for the serendipitous discovery of penicillin. The lesson Bourne preaches from this text is the absolute necessity of basic research, the stock on which we draw for applications. The stock can dry up and wither. We must as a culture again learn the usefulness of useless knowledge. It is as nature to culture—the resource for triumph over sickness, over others; the ground of our birth as men.

Wiseman then leads us to the payoff of *Primate.* A van draws up to the rear of the laboratory to remove its product, a rhesus in a plastic restraining chair. The monkey is driven to a military airfield and placed with a mechanical lift into the belly of a large plane. The animal shares the space with several high-altitude-suited men. But the monkey is placed in a black box and watched by the men on a TV screen for the rest of the film. The channels of the monkey's brain are recorded by other equipment. The plane carrying this cargo is watched on a radar screen at the airfield. We, in turn, see all the screens: the airplane diving and rising; the monitors of the radar; the humans floating in the belly of the plane, connected by thick cables to life-sustaining equipment; the monkey staring out at us from its box; the credits of the movie itself. The film is over. Primates are in space, connected to earth by the technology of communication, birthing streams of information, transforming nature into culture according to an everyday logic of domination in a scientific division of labor. It is only a myth, of course, an origin story among others. A payload.

NOTES

1. Conference on the Procurement and Production of Rhesus Monkeys, Institute of Animal Resources, National Research Council, June 7, 1955, and September 22, 1955, Clarence Ray Carpenter Papers (hereafter cited as CRC Papers), "Nonhuman Primate Procurement and Production" folder, Pennsylvania State Room, Fred Lewis Pattee Library, Pennsylvania State University, University Park. The Indian authorities lifted the ban after 4 months under a new agreement with the U.S. that set standards for use and shipment. Specifically forbidden were any uses of monkeys in atomic and space research. The agreement was regularly violated from at least 1966, resulting in a new Indian embargo in 1978 in response to public discovery that the animals were subjects in neutron bomb experiments. International Primate Protection Leage, *Newsletter,* Summerville, S.C., April 1978, pp. 2–6.

2. Hamilton Cravens, *The Triumph of Evolution: American Scientists and the Heredity Environment Controversy, 1900–1941* (Philadelphia: University of Pennsylvania Press, 1978), discusses the resolution of the heredity-environment controversy in America by the 1930s in terms of an "interactionist paradigm" for nature and culture.

3. This logic is thoroughly explored in Erik Rusten Hogness, "Herbert Spencer and the Ideology of Beneficent Necessity: The Naturalization of Capitalist Society and Middle-Class Liberalism in Spencer's Early Work," M.A. thesis, The Johns Hopkins University, 1979.

4. Bruno Latour and Steve Woolgar, *Laboratory Life: The Social Construction of Scientific Facts* (Beverly Hills and London: Sage, 1979), develop this theme in detail for the production of a particular fact, thyrotropin releasing factor, TRF, in Roger Guillemin's laboratory at the Salk Institute.

5. Paul Weiss, Remarks of the Chairman of the Division of Biology and Agriculture, "Non-Human Primate Committee Report to the Institute of Animal Resources," September 22, 1955, p. 1, CRC Papers, "Nonhuman Primate Procurement and Production" folder.

6. Weiss, "Non-Human Primate Committee Report," p. 2.

7. Weiss, "Non-Human Primate Committee Report," p. 3.

8. Robert Means Yerkes, *Chimpanzees: A Laboratory Colony* (New Haven: Yale University Press, 1943), pp. 9–10. For a full discussion of Yerkes's role in primate studies, see Donna Haraway, "Sex, Primates, and Human Engineering: The Laboratory as Pilot Plant, 1924–42," paper presented at Conference on the Study of Fertility, American Academy of Arts and Sciences, May 1978, Boston. MS available from the author.

9. R[obert] M. Yerkes, "Yale Laboratories of Comparative Psychobiology," *Comparative Psychology Monographs* 8, no. 3 (1932): 1–33, and "Yale Laboratories of Primate Biology, Inc." *Science* 82 (1935): 618–20; Geoffrey H. Bourne, ed., *Progress in Ape Research* (New York: Academic Press, 1977); H. C. Bingham, "Gorillas in a Native Habitat," *Carnegie Institute Publications* 426 (1932): 1–66; H. W. Nissen, "A Field Study of the Chimpanzee," *Comparative Psychology Monographs* 8, no. 8 (1931): 1–122; and C. R. Carpenter, "A Field Study of the Behavior and Social Relations of Howling Monkeys," *Comparative Psychology Monographs* 10, no. 2 (1934): 1–168, reprinted in C. R. Carpenter, *Naturalistic Behavior of Nonhuman Primates* (University Park: Pennsylvania State University Press, 1964), pp 3–92, this collection of Carpenter papers is hereafter referred to as NB.

10. Weiss, "Non-Human Primate Committee Report," p. 2.

11. U.S., Department of Health, Education, and Welfare, Public Health Service, NIH, *National Institutes of Health Primate Research Centers: A Major Scientific Resource,* December 1971, and *Regional Primate Research Centers: The Creation of a Program,* 1968, 67 pp. About 67,000 nonhuman primates per year were used in medical research in 1969. The decline from the high during the 1950s reflected changes in the production of polio vaccine.

12. Weiss, "Non-Human Primate Committee Report," p. 1.

13. For a visual treatment of the theme of biological research as large-scale, highly automated industrial production, see Frederick Wiseman, *Primate,* available from Xippora Films, Boston. *Primate* is a documentary based on the Yerkes Regional Primate Research Center. See sec. 5.

14. Donna Haraway, "The Biological Enterprise: Sex, Mind, and Profit from Human Engineering to Sociobiology," *Radical History Review* 20 (1979): 206–37, and "The High Cost of Information in Post–World War II Evolutionary Biology: Ergonomics, Semiotics, and the Sociobiology of Communication Systems," *Philsophical Forum* 13, nos. 2–3 (winter–spring 1981–82): 244–78.

15. The debate over the proper taxonomic status of monkeys and apes, and the bearing of that vexed question on the human place in nature, is considered by Kristen Zacharias in her Ph.D. dissertation, "The Construction of a Primate Order: Taxonomy and Comparative Anatomy in Establishing the Human Place in Nature, 1735–1916," The Johns Hopkins University, 1980. Linnaeus first created the order Primates.

16. C. R. Carpenter, "A Study of Sex Behavior of the Common Pigeon with Special Emphasis on the Monogamic Tendencies of Mated Pairs," M.A. thesis, Duke University, 1929, "Psychobiological Studies of Social Behavior in Apes: I. The Effect of Complete and Incomplete Gonadectomy on Primary Sexual Activity of the Male Pigeon," *Journal of Comparative Psychology* 16 (1933): 25–27, and "Psychobiological Studies of Social Behavior in Apes: The Effect of Complete and Incomplete Gonadectomy on Secondary Sexual Activity with Histological Studies," *Journal of Comparative Psychology* 16 (1933): 59–97.

17. Sophie D. Aberle and George W. Corner, *Twenty-five Years of Sex Research: History of the National Research Council Committee for Research in Problems of Sex, 1922–47* (Philadelphia: W. B. Saunders, 1953); and National Academy of Sciences Archives, Committee for Research in Problems of Sex Papers, hereinafter cited as CRPS Papers, Calvin Stone file.

18. R[obert] M. Yerkes, "Social Dominance and Sexual Status in the Chimpanzee," *Quarterly Review of Biology* 14 (1939): 115–36; Solly Zuckerman, *The Social Life of Monkeys and Apes* (London: Routledge & Kegan Paul, 1932); and G. S. Miller, "The Primate Basis of Human Sexual Behavior," *Quarterly Review of Biology* 6 (1931): 379–410. For an understanding of why sex was constituted as such a fundamental organon of power over life conceived as productive systems, see Michel Foucault, *A History of Sexuality* (New York: Pantheon, 1974). Functionalism in its various forms has been the scientific understanding of such fruitful life.

19. This focus neglects critical aspects of the story of postwar primatology, especially in relation to the dissenters from cybernetic functionalism. That story must be told in another context. Personal communication from Sherwood Washburn, June 17, 1979, and interview with Sherwood Washburn, Berkeley, Calif., August 2, 1979. Carpenter and Washburn represented different approaches to the concept of social structure.

20. Charles F. Hoban and E. B. Van Ormer, *Instructional Film Research, 1918–50* (New York: Arno Press Reprint, 1970). This comprehensive bibliography was prepared by the Pennsylvania State Instructional Film Research Program, sponsored by the Departments of the Army and Navy.

21. J. A. Hrones, Foreword to *Systems Theory and Biology,* ed. M. D. Mesarovic (New York: Springer-Verlag, 1968), p. vii, points out that "in the early 1920s a further development of the dynamics of systems led to the first automatic steering system for ocean-going vessels. However the greatest single thrust [thrust!] to the development of systems probably occurred in the 1930s with the growth of continental and intercontinental communication systems." The next step in control systems was operations research in World War II. Sherwood Washburn strongly argues that changes in primate research before and after World War II can be largely accounted for by the airplane. Personal interview, July 1977, and correspondence, July 12, 1979. The point is correct in theory as well as practice. The precybernetic physiologies of self-regulating systems of interest in this essay are represented by the physiological laboratories at Harvard of Walter Cannon and Lawrence J. Henderson.

22. Carpenter to Yerkes, February 18, 1931, Robert M. Yerkes Papers (hereafter cited as RMY Papers), folder 82, Dept. of Manuscripts and Archives, Sterling Memorial Library, Yale University, New Haven; and letters of recommendation from Stone, Miles, Terman, and McDougall, National Research Fellowship records, National Academy of Sciences Archives, Washington, D.C.

23. Carpenter application to National Research Fellowship Board, "Social Relations in Animals," RMY Papers, folder 82.

24. Yerkes to Carpenter, March 10, 1931; Yerkes to National Research Fellowship Board, March 28, 1931; and Yerkes to Walter Miles, Yale Institute of Human Relations, seeking Social Science Research Council support for Carpenter, March 10, 1931, RMY Papers, folder 82.

25. Carpenter application to National Research Fellowship Board, "Social Reactions in Animals," March 10, 1931, RMY Papers, folder 82. The other sketches in the application are titled "The Effect of Castration on the Monkey" and "An Experiment in Group Learning in Chimpanzees."

26. Yerkes to Thomas Barbour, Museum of Comparative Zoology and Institute for Research in Tropical America, Barro Colorado Island Biological Laboratory, November 14, 1931, RMY Papers.

27. From Chapman to Yerkes, quoted in Robert Yerkes's Foreword to Carpenter, "Behavior and Social Relations of Howling Monkeys." Barro Colorado Island Biological Laboratories have an interesting story in relation to the "workshop" system of production of science. It was part of the Institute for Research in Tropical America, set up under National Research Council auspices in 1923. Washington, D.C., National Academy of Sciences Archives, Institute for Research in Tropical America. Thomas Barbour, *A Naturalist in Cuba* (Boston: Little, Brown, 1945). F. M. Chapman, *My Tropical Air Castle* (New York: Appleton, 1929). In chap. 11, "The Monkeys," Chapman demonstrated close knowledge of howlers. David Fairchild, "Barro Colorado Island Laboratory," *Journal of Heredity* 15 (1924): 99–112. The apparent greater role of entrepreneurial individuals characterizes the workshop theme.

28. Reaching back into the eighteenth century, physical anthropology is another parent of primatology. It was not neglected by Carpenter, who read and cited G. Elliot-Smith, Adolph Schultz, J. P. Wood-Jones, Arthur Keith, and Harold Coolidge. But Carpenter did not make any important contributions to primatology from the point of view of physical anthropology, nor did he appear to understand or appreciate the fundamental revolution in primate studies issuing from the work of Sherwood Washburn and his students since the 1950s. Carpenter was highly critical of the Washburn approach embodied in the 1962–63 National Institutes of Mental Health–sponsored Primate Project at the Center for Advanced Study in the Behavioral Sciences in Stanford, Calif. Carpenter to Preston Cutler, March 9, 1964 and May 19, 1964, on publication of papers from the year's seminar, CRC Papers. Irven DeVore, ed., *Primate Behavior: Field Studies on Monkeys and Apes* (New York: Holt, Rinehart & Winston, 1965). Donna Haraway, "Animal Sociology and a Natural Economy of the Body Politic, II: The Past Is the Contested Zone," *Signs,* 4 (1978): 37–60.

29. Carpenter to Yerkes, March 4, 1931, RMY Papers.

30. Charles Otis Whitman, "The Behavior of Pigeons," *Carnegie Institution of Washington Publications* 257 (1919): 1–161, ed. (posthumously) H. Carr; and Wallace Craig, "The Voice of Pigeons Regarded as Means of Social Control," *American Journal of Sociology* 19 (1908): 29–80, and "Appetites and Aversions as Constituents of Instincts," *Biological Bulletin* 34 (1918): 91–107. A comparison and contrast of Lorenz's motor action patterns and Carpenter's would be fruitful. In the 1930s they both performed careful anatomical investigations of movement and posture in free animals and related their studies to coordination of behavior, especially sexual behavior, in forming organic societies. They were interested in motor patterns and vocalizations as forms of communication integrating individuals into a larger functional whole. The behavioral forms and the social forms, like other biological structures, were subject to evolution. Margaret Nice, an American student of sparrows, called Carpenter's attention in 1935 to Lorenz's 1931 and 1932 papers. Nice and Lorenz had been in Oxford at the International Ornithological Conference. Ernst Mayr told Nice to consult Carpenter about her work. Nice to Carpenter, March 16, 1935, and March 29, 1935, CRC Papers. See also M. M. Nice, *Studies on the Life History of the Song Sparrow,* vol. 1: *A Population Study of the Song Sparrow, Transactions of the Linnaean Society,* vol. 4, no. 6 New York, 1937), 247 pp., and "The Theory of Territorialism and Its Development," in F. M. Chapman and T. S. Palmer, eds., *Fifty Years of Progress in American Ornithology, 1883–1933* (Lancaster, Pa.: American Ornithologists Union, 1933), pp. 89–100. Between 1930 and 1970, Lorenz transformed his ethological theory of instinct according to the same logic that this essay argues generally affected biology, leading from functional morphology and physiology to cybernetics. Robert J. Richards, "The Innate

and the Learned: The Evolution of Konrad Lorenz's Theory of Instinct," *Philosophy of Social Science* 4 (1974): 111–33, p. 127. Lorenz's model of instinctive action changed from a hydraulic machine to an information machine. Important issues in the history of instinct theory may be followed in Robert J. Richards, "Lloyd Morgan's Theory of Instinct: From Darwinism to neo-Darwinism," *Journal of the History of Behavioral Sciences* 13 (1977): 12–32. Carpenter, a student of William McDougall, was a complete mix of neo-Lamarckian and neo-Darwinian strands in the 1930s. But these issues go beyond the bounds of this paper.

31. Carpenter wrote a review of the concept of territoriality for a volume resulting from a deliberate effort to incorporate comparative psychology into the new evolutionary doctrine widely called the New Synthesis. Two conferences were sponsored by the American Psychological Association and the Society for the Study of Evolution, in 1955 and 1956. Primate studies were well represented: Carpenter, Harry Harlow, Henry Nissen, and Sherwood Washburn. See Anne Row and George Gaylord Simpson, eds., *Behavior and Evolution* (New Haven: Yale University Press, 1958), especially "Territoriality: A Review of Concepts and Problems," pp. 224–50. In his paper for the first conference, "Barriers and Approaches to a Community of Understanding of the Evolution of Behavior" (April 4, 1955, CRC Papers), Carpenter was unimpressed by the implications for comparative psychology of the new evolutionary work. He equated slow progress in science with infrequency of geniuses. His enthusiasm for systems theories and new semantic, logical, and mathematical methods was large; but he seemed not to understand the theoretical issues in the new evolutionary synthesis. His final paper on territory for the second conference reflects this confusion. He believed that territoriality was important, but he did not have a consistent point of view, confusing organismic physiological functionalism and the newer evolutionary population biology. I think his howler monkey paper draws heavily on the ornithological work on territory without really understanding it. According to Washburn, for whom the new synthesis was the foundation of his work, Ernst Mayr, a consummate bird expert, was exasperated by Carpenter's inability to understand territoriality in the context of the purposes of the conference. Washburn to Haraway, July 4, 1979. H. E. Howard's *Territory in Bird Life* (London: Murray, 1920) and Ernst Mayr's translation of Altum's 1868 work, "Bernard Altum and the Territory Theory," *Proceedings of the Linnaen Society,* nos. 45–46 (New York, 1935), indicate the role of ornithology in the early theory of territoriality.

32. G. Evelyn Hutchinson, *An Introduction to Population Ecology* (New Haven: Yale University Press, 1978) makes clear how important to ecology the concepts of space have been. See esp. chap. 5. Space is an abstract concept, not an immediately present physical object. Ernst Mayr, "Epilogue on American Ornithology," in Erwin Stresemann, ed., *Ornithology: From Aristotle to the Present* (Cambridge: Harvard University Press, 1975), pp. 365–96, provides rich material for following further the connections of ornithology and primate studies. Figures who will appear again in the Carpenter study are John Emlen, Nicholas Collias, G. K. Noble, and Frank Chapman. Emlen, a graduate student of Arthur Allen, of Cornell, is a major sociobiologist who has done basic work on communication and territory. George Schaller, who studied the mountain gorilla in the early 1960s, was Emlen's graduate student. Allen was the first university professor of ornithology in the U.S. Arthur A. Allen, "Cornell's Laboratory of Ornithology," *Living Bird* 1 (1962): 7–36. Collias collaborated with an important figure in population biology and resource conservation efforts of howlers and rhesus, Charles Southwick. Collias was one of several channels of European ethology into primatology. Nicholas E. Collias, "Social Life and the Individual Among Vertebrate Animals," *Annals of the New York Academy of Sciences* 51 (1950): 1076–92, and N[icholas] E. Collias and C. H. Southwick, "A Field Study of Population Density and Social Organization in Howling Monkeys," *Proceedings of the American Philosophical Society* 96 (1952): 143–56. Robert Hinde and John Hurrell Crook are two additional central people in the interweaving of bird and primate work. All of these people have roles in the transformation of animal behavior into a communications science called sociobiology. G. K. Noble, a Committee for Research in Problems of Sex grantee in the 1930s, was followed at the American Museum of Natural History by Frank Beach, another central figure in the history of instinct and reproductive and neural physiology. Working on rodents, fish, amphibians, and birds, Noble was important in the role of dominance theory in the physiology of animal coordination. G. K. Noble, "The Role of Dominance in the

Social Life of Birds," *Auk* 56 (1939): 263–73. Finally, the history of ornithology, with its late university professionalization among the biologies and long connections with museums and field expeditions, is helpful for following the transformation of biological research from "workshop" to "industrial" conditions. Primatology shares some of the same dates in this process as ornithology.

33. Mayr, "Epilogue on American Ornithology."

34. On the history of human population statistics and policy planning based on forecasts, see William Ascher, *Forecasting, An Appraisal for Policy-Makers and Planners* (Baltimore: Johns Hopkins University Press, 1978). Raymond Pearl is the key figure for the connection of study of human and animal populations alike as physiological objects, for which variations in growth form are biological phenomena subject to medical management. For a succinct picture of Pearl's approach to populations, see his "Some aspects of the Biology of Human Populations," in E. V. Cowdry, ed., *Human Biology and Racial Welfare* (New York: Hoeber, 1930). This book is a powerful example of the organicism and physiological functionalism of biology in the 1920s and 1930s. It was intended for men of business or science who wished an overview of human biology critical to a confused human civilization searching for confidence and stability. See the Introduction, by Edwin Embree, the officer of the Rockefeller Foundation responsible for the first grant to Robert Yerkes for chimpanzee studies at Yale's new Institute of Psychology in 1924. Raymond Pearl, with L. J. Reed, rediscovered the famous logistic equation to describe population growth: "On the Rate of Growth of the Population of the United States since 1790 and Its Mathematical Representation," *Proceedings of the National Academy of Sciences* 6 (1920): 275–88. See Hutchinson, *Introduction to Population Ecology,* pp. 1–40. Pearl's complex career included work on domestic fowl populations when he was at the biological research station in Maine before World War I. Mathematical population modeling has become a principal operation in a postwar biology based on cybernetic machines and communication theories. Some of the theoretical foundation for this development was laid by the longtime employee of the Metropolitan Life Insurance Company Alfred J. Lotka, who worked for a year in the mid-1920s at The Johns Hopkins University School of Hygiene and Public Health in association with Pearl. The result was the production of the long-gestating A. J. Lotka, *Elements of Mathematical Biology* (Baltimore: Williams & Wilkins, 1925; reprinted as *Elements of Mathematical Biology,* Dover, 1956). Lotka is useful in studying the interconnections of insurance interests and biological theory. Political and natural economy have a long history together; economics and bionomics are both management sciences based on production systems. Francois Jacob, *The Logic of Life,* trans. Betty Spillman (New York: Pantheon, 1974).

35. W. C. Allee, *Cooperation among Animals* (New York: Schuman, 1951), p. 136; and Thorlief Schjelderup-Ebbe, "Beitrage zur Sozialpsychologie des Haushuhns," *Zeitschrift fur Psychologie* 88 (1922): 225–52, and "Social Behavior of Birds," in C. Murchison, ed., *A Handbook of Social Psychology* (Worcester, Mass.: Clark University press, 1935), pp. 947–72.

36. Meredith Crawford, "The Social Psychology of Vertebrates," *Psychological Bulletin* 36 (1939): 407–46, pp. 418–19. This historical review article on studies since 1900 of the interactive behavior of organisms dedicated more space to dominance than to any other issue, 10 pages.

37. Joseph A. Caron, "La theorie de la cooperation animale dans l'ecologie de W. C. Allee: Analyse du double registre d'un discours." M.S. thesis at L'Institut d'histoire et de sociopolitique des sciences, Université de Montréal, 1977; and W. C. Allee, *Animal Cooperation, A Study in General Sociology* (Chicago: University of Chicago Press, 1931), "Social Dominance and Subordination among Vertebrates," *Biological Symposium* 8 (1942): 139–62; and "Animal Sociology," *Encyclopaedia Britannica,* 1954: pp. 914A–916.

38. W. C. Allee, A. E. Emerson, O. Park, and K. P. Schmidt, *Principles of Animal Ecology* (Philadelphia: W. B. Saunders, 1949), p. 9. The introduction to this book is an important historical essay. See esp. pp. 55–62 for a discussion of the years 1921–30.

39. Allee et al., *Principles of Animal Ecology,* p. 6.

40. Robert P. McIntosh, "Ecology since 1900," in B. J. Taylor and T. J. White, eds., *Issues and Ideas in America* (Norman: University of Oklahoma Press, 1977), pp. 353–72; Victor E. Shelford, *Animal Communities in Temperate America* (Chicago: University of Chicago Press, 1913); and F. E. Clements and V[ictor] E. Shelford, *Bioecology* (New York:

Wiley, 1939). Transformation of ecology from a science based on communities to one based on cybernetic systems is discussed in reference to the work of E. P. Odum by Michael Farley, "Formations et transformations de la synthese écologique aux Etats-Unis, 1949–1971," M.S. thesis, L'Institut d'histoire et de sociopolitique des sciences, Université de Montréal, 1977. Odum speaks for himself in "The Emergence of Ecology as a New Integrative Discipline," *Science* 195 (1977): 1289–93.

41. Garland Allen, *Life Science in the Twentieth Century* (New York: Wiley, 1975), pp. 88–112; John Parascondola, "Organismic Concepts in the Thought of L. J. Henderson," *Journal of the History of Biology* 4 (1971): 63–114; Cynthia Russett, *The Concept of Equilibrium in American Social Thought* (New Haven: Yale University Press, 1966); Conrad Hal Waddington, "Morphogenetic Fields," *Science Congress* 114 (1934): 336–46; Donna Haraway, *Crystals, Fabrics, and Fields: Organismic Metaphors in 20th Century Developmental Biology* (New Haven: Yale University Press, 1976); and Joseph Needham, *Order and Life* (New Haven: Yale University Press, 1936). Paul Weiss, the chairperson of the meeting to counter the effects of the Indian rhesus embargo discussed in sec. 1, was a major organismic developmental biologist of the 1930s, who helped effect a transition to a cybernetic systems functionalism after World War II. Carpenter named his three scientific models, all organismic, physiological functionalists: Yerkes, Chapman, and William Morton Wheeler. Carpenter to Yerkes, December 14, 1937, RMY Papers, folder 84. W. M. Wheeler, *The Social Insects: Their Origin and Evolution* (New York: Harcourt Brace, 1928), and *Essays in Philosophical Biology* (Cambridge: Harvard University Press, 1939). Wheeler worked often on Barro Colorado.

42. C. M. Child, "Biological Foundations of Social Integration," *Publications of the American Sociological Society* 221 (1928): 26–42, pp. 32, 42.

43. C. M. Child, "Social Integration as a Biological Process," *American Naturalist* 74 (1940): 389–97, pp. 391, 394, 395, and 397. These sorts of principles are not restricted to texts that directly compare society and protoplasm. C. M. Child, *Physiological Foundations of Behavior* (New York: Holt, 1924). Child was at the University of Chicago with Allee.

44. A. E. Emerson, "Social Coordination and the Superorganism," *American Midland Naturalist* 21 (1939): 182–209; and "Dynamic Homeostasis: A Unifying Principle in Organic, Social, and Ethical Evolution," *Scientific Monthly* 78 (1954): 67–85. This is the same Emerson who co-authored *Principles of Animal Ecology* in 1949 with Allee. It is important to note that organic materialism (or physiological functionalism) of complex wholes ordered by homeostatic mechanisms, chiefly dominance, was entirely compatible with the autonomy of the biological and social sciences established by the late 1930s in America. Cravens, *Triumph of Evolution*. The superorganism point of view was transformed with cybernetic functionalism and its legitimate offspring, sociobiology. Haraway, "Biological Enterprise."

45. Yerkes to Carpenter, October 1, 1931, RMY Papers, folder 82. In fact Chapman was not a "doctor"; his high professional status without a Ph.D. was still possible in ornithology in the 1930s.

46. Carpenter's opportunity was not only remarkable in terms of scientific understanding of primates; it was perceived as a very desirable and much appreciated advance for a man from a poor family in Cherryville, N.C. Carpenter's parents wrote a moving letter to Yerkes on December 22, 1931, thanking him for his many kindnesses and expressing their worry about their son heading for a foreign and strange place so far from home. C. E. and G. L. Carpenter to Yerkes, RMY Papers, folder 82.

47. Carpenter to Yerkes, December 31, RMY Papers, folder 82.

48. Chapman to Yerkes, February 18, 1932, RMY Papers, folder 82.

49. Chapman to Yerkes, January 29, 1932, RMY Papers, folder 82. From the beginning Carpenter photographed his howlers, following the lead of Chapman, who had introduced photography into his bird work in 1908. Cf. Carpenter to Yerkes, February 5, 1932, RMY Papers; and Mayr, "Epilogue on American Ornithology," pp. 391–92. Many of Carpenter's letters to Yerkes concerned photographic equipment, e.g., in Carpenter to Yerkes, May 13, 1932, Carpenter is consulting his friend at General Electric about field apparatus. Eventually, Carpenter's concern with films, which he considered totally objective, occupied his major attention. The filming experience is intimately related to the theme of a science of communications.

50. Robert Yerkes and Ada Yerkes, *The Great Apes* (New Haven: Yale University Press, 1929).

51. Zuckerman, *Social Life of Monkeys and Apes.* Yerkes to Carpenter, January 18, 1932, RMY Papers, Yerkes considered this book "will be most valuable"; but by March 25, 1932, he cautioned Carpenter against Zuckerman's overgeneralizations, especially on apes. His work on baboons and other monkeys was still regarded as reliable. Zuckerman's views related to debates about the origin of human society and were developed as an answer to Bronislaw Malinowski's ideas on the origin of the family on the basis of unique human female physiology (menstruation) and of the original cultural institution (fatherhood).

52. Carpenter to Yerkes, March 26, 1932, RMY Papers.

53. Carpenter to Thomas Barbour, February 9, 1933, RMY papers; and Carpenter field notes: "Field Studies at Camp La Vaca, the Clark Expedition," pp. 47–48, 54–56, and 58–61, CRC Papers.

54. Carpenter to Yerkes, May 13, 1932, RMY Papers.

55. C. R. Carpenter, "Behavior of red spider monkeys in Panama," *Journal of Mammology* 16 (1935): 171–80. Carpenter to Yerkes, July 8, 1932, and March 5, 1933, RMY Papers. Carpenter field notes for 1932, "Field Studies at Camp La Vaca, the Clark Expedition," on spiker monkeys, pp. 29–36 and 50; and "Book 1," on howlers, p. 145, CRC Papers.

56. Zuckerman's thesis that primates did not have a breeding season and that female sexual receptivity was the glue of nonhuman primate society influenced workers for decades, despite the lack of field verification. Laboratory macaques seemed to confirm the absence of defined breeding seasons. But more fundamentally, sex was *expected* to function as the danger-fraught organic base of society, contained at the human level by cultural control (kinship systems) and ordered at the natural level by dominance hierarchies. Miller, "Primate Basis of Human Sexual Behavior;" S. L. Washburn, "Human Behavior and the Behavior of Other Animals," *American Psychologist* 33 (1978): 405–18, and Jane Lancaster and Richard Lee, "The Animal Reproductive Cycle in Monkeys and Apes," in DeVore, ed., *Primate Behavior.* The expectation was so strong that Carpenter considered his own recorded observations that Cayo Santiago rhesus *did* have a well-marked breeding season to be very questionable. Carpenter, remarks, Conference on the Procurement and Production of Rhesus Monkeys, June 7, 1955, pp. 41–42, CRC Papers.

57. Hartman worked in one of the first primate laboratories, established by George Washington Corner in 1923 to study menstrual physiology. Human social and physiological medical problems were always in view. Carl Hartman, "Studies on Reproduction in the Monkey and Their Bearing in Gynecology and Anthropology," *Bulletin of the Association for the Study of Internal Secretions* 25 (1932): 670–82. This laboratory had major funding from the Yerkes-chaired Committee for Research in Problems of Sex. Aberle and Corner, *Twenty-five Years of Sex Research,* p. 119.

58. Carpenter to Stone, April 5, 1934, CRC Papers, box 2, 1933–36, Correspondence. Looking for a job, Carpenter was complaining to Stone that positions were scarce and he would rather go back to the comparative psychology laboratory than do more field work. "Such research is extremely difficult, uncertain and in many ways unsatisfactory. Furthermore, after one has done field work for *15* years he probably would not be fitted for anything else." The uncertain foundation of primate studies until the international explosion of interest beginning in the mid-1950s must be emphasized. As late as 1953, Carpenter thought he would never write another primate paper because, from the point of view of comparative psychology, the field was moribund. The reawakening in the mid and late fifties, international in scope, occurred in altered material and theoretical circumstances. Carpenter to Yerkes, June 16, 1952, Carpenter to Yerkes, June 26, 1953, and Yerkes to Carpenter, July 2, 1953, RMY Papers, folder 84. Carpenter's "last paper" was presented to a French symposium on primate behavior held in 1950. Retrospectively, this conference was a sign of the coming reawakening. C. R. Carpenter, "Social Behavior of Nonhuman Primates," *Colloques Internationaux du Centre Nationale de la Recherche Scientifique* 34 (1952): 227–46. Yerkes's lab was permanently established, but ran into major funding problems in the Depression, not only because of general economic trouble, but primarily because his brand of psychobiology and physiological functionalism progressively lost favor with Warren Weaver of the Rockefeller Foundation. *Annual Reports of Yale Laboratories for Primate Biology,* 1935–42, RMY papers. Carpenter's howler paper got a quiet reception.

59. C. R. Carpenter, "The Origins and Present Status of the Breeding Colony of Macaques *(Macaca mulatta)* on Santiago Island, Puerto Rico," Athens, Ga., June 1971, pp. 2–3, MS in CRC Papers.

60. Hartman to Carpenter, January 17, 1966, CRC Papers, "Primate Centers, Puerto Rico" folder. Cayo Santiago's troubles continued well into the 1960s, but that story must await another article.

61. Wislocki to Carpenter, February 25, 1935, CRC Papers, box 2, 1933–36, Correspondence.

62. Wislocki to Carpenter, June 21, and July 19, 1935, CRC Papers, Box 2, 1933–36, Correspondence. Nothing came of the plan.

63. Carpenter to Wislocki, May 2, and May 18, 1936; and Wislocki to Carpenter, May 21, 1936, CRC Papers. Yerkes's support for A.P.E. on the CRPS could not be assumed. After Carpenter's postdoctoral period, he did not go out of his way for Carpenter on the committee. Yerkes to Carpenter, May 26, 1934; June 25, 1934; and October 4, 1934, RMY Papers. Yerkes was having his own funding problems at that time. He directed Carpenter's attention to other sources.

64. Carpenter field notes: b.1, pp. 30, 34, 50, 45, and 81; b.2, pp. 16 and 20, CRC Papers.

65. Many of Carpenter's field notes concerned drive behaviors in relation to learning situations, especially for infants nursing and playing. He labeled several sections of his notebooks to record classical data for comparative psychologists: multiple-choice situations, intelligent action, instrumentation, discrimination and percept, imitation and suggestibility, social facilitation, behavioral adaptation, behavior indicative of memory, examples of delayed reaction.

66. Carpenter, "Behavior and Social Relations of Howling Monkeys."

67. Carpenter's films did the same thing. Environment, group, animal elements, motions, calls, and so on were all presented so as to convince the viewer of the essential absence of the filmer or scientist. Unsurprisingly, this banal stylistic affiar is ideologically important in lending credence to generalizations about human primates based on observations of nonhuman primates. The psychological conviction is that the direction of knowledge is from animal to people, not the reverse. Some of his primate films are: *Behavior of the Macaques of Japan,* 1969; *Howler Monkeys of Barro Colorado Island; Social Behavior of the Rhesus Monkey;* and *Characteristics of Gibbon Behavior.* All are listed in the Psychology Cinema Register and are available from the Audio-Visual Center, Pennsylvania State University.

68. *NB,* p. 35.

69. *NB,* p. 35.

70. *NB,* p. 46.

71. *NB,* p. 47.

72. *NB,* pp. 47–48.

73. *NB,* p. 66.

74. General Statement of Plans for Study, October 28, 1932, p. 3, RMY Papers. Yerkes's letter of support stressed the sociological relevance of the project. Yerkes to Guggenheim Memorial Foundation, October 3, 1932, RMY Papers.

75. C. R. Carpenter, "Specific Plans for a Psychobiological and Comparative Sociological Study of Playtrrhines," November 9, 1932, p. 5, RMY Papers.

76. The absence of much dominance behavior among the howler was a matter for constant amazement and comment. Noting that primate life does not seem to be the ruthless struggle often depicted, the Yale University press release on the howler monograph emphasized that "howlers lead a communal type of life, with no individual dominating the group." Yale News Release, June 11, 1934, RMY Papers. This virtue in natural cooperation was more difficult for more complex societies. Harvard anthropologist E. A. Hooton, *Man's Poor Relations* (New York: Doubleday, 1942) highlighted the communistic howlers of Carpenter and the despotic hamadryas of Zuckerman. In popular primate literature cautionary moral tales abound.

77. Christopher Lasch, *Haven in a Heartless World* (New York: Basic Books, 1977), esp. chap. 2, "Sociological Study of the Family in the Twenties and Thirties." Lasch stresses the definition of the key object of "autonomous" social science, namely *society,* as the totality of social roles. At bottom, then, society was a problem in communication, from face-to-face encounters to ever-widening networks. Lasch illuminates the organismic and

positivist frameworks of the Chicago school of urban sociology, with their ideas on sentiment and sympathy as the social glue. These are the very sociologists recognized for their adoption of the culture concept and *rejection* of Spencerian relations of biological and social science!

78. One consequence of this congruity has been the continuing ideological power of popularizations of animal sociology. This power does not derive from biological reductionism, but from the mind-numbing repetition of shared logics of domination. This literature teaches there is no crack in the biological and social body theorized as a problem in the control of communication. See n. 191 on Robert Ardrey and sec. 5, "Primate."

79. C. R. Carpenter, "An Observational Study of Two Captive Mountain Gorillas *(Gorilla beringii),*" *Human Biology* 9 (1937): 175–96. Yerkes was reluctant to recommend Carpenter for this study because Harold C. Bingham, another Yerkes associate, who had conducted the first modern field gorilla study under the Yale auspices and who had studied the psychosexual development of 4 young chimpanzees, was also interested in doing the observations in San Diego. Bingham withdrew from consideration, so Carpenter did the work. H. C. Bingham, "Sex Development in Apes," *Comparative Psychology Monographs* 5 (1928): 1–165, and "Gorillas in a Native Habitat"; Carpenter to Yerkes, February 5, 1934; Yerkes to Carpenter, February 10, 1934; and Carpenter to Yerkes, February 13, 1934, RMY Papers.

80. Aberle and Corner, *Twenty-five Years of Sex Research,* pp. 45–51; and "Projects, Resarch Centers, Psychobiology of Sex in Man, Boston Center, 1923" folder, and "Projects, Research Centers, Marital Research, G. V. Hamilton, 1923–24" folder, CRPS Papers.

81. E. L. Kelley, "A preliminary Report on Psychological Factors in Assortative Mating," *Psychological Bulletin* 34 (1937): 749, and "A 36 Point Personality Rating Scale," *Journal of Psychology* 9 (1940): 97–102; and L. M. Terman and C. C. Miles, *Sex and Personality: Studies in Masculinity and Femininity* (New York: McGraw-Hill, 1936).

82. Carpenter to Kelley, April 21, 1934, CRC Papers. Carpenter expected to conduct this work through the auspices of the Institute of Human Relations at Yale.

83. Carpenter to Yerkes, February 22, 1936, RMY papers; and Carpenter field notebook for "B.C.I.—Bard College Exp., 1935–36," p. 66, CRC Papers.

84. Carpenter to Poffenberger, of the Columbia Council for Research in Social Science, March 27, 1936, CRC Papers; and C. R. Carpenter, "An Investigation of Masculine Compatibility," submitted to the Columbia Council for Research in Social Science, April 1, 1936, pp. 2 and 4, RMY Papers, folder 84.

85. C. R. Carpenter, "A Field Study in Siam of the Behavior and Social Relations of the Gibbon *(Hylobates lar),*" in NB, p. 161.

86. Carl E. Akeley, *In Brightest Africa* (London: Heinemann, 1924).

87. Coolidge to Yerkes, June 6, 1936, RMY Papers. Coolidge urged Yerkes's complete confidentiality on these reasons for the expedition. The public press announcements of the expedition stressed "Party to Trace Human Traits in Asiatic Apes" *(New York Herald Tribune*, December 16, 1936, p. 25) and "Studying the Key Ape: The Gibbon in Siam May Unlock Secrets of Man's Past" *(New York Times,* December 2, 1936). The evolution of marriage and posture were highlighted in these articles. RMY papers, Carpenter correspondence, folder 84, has important documentation of the funding and organizational aspects of the expedition. Carpenter's research was largely covered by $2,300 from the Columbia Council for Research in Social Science, with $1,000 added by Coolidge.

88. But it is important to remember how much such "individual" research relied on a social foundation. For example, in Carpenter's case in Panama, research was only possible with the United Fruit Company travel accommodations, the National Academy of Sciences–National Research Council role in founding a laboratory, the Canal Zone authority in setting aside hilltop land made into an island by filling Gatun Lake, Yerkes's Rockefeller Foundation–funded Yale Laboratories of Comparative Psychobiology, National Research Fellowships developed on the basis of science's role in war and industry, and the "private" wealth of individuals at B.C.I., like Thomas Barbour. The scientific study of "coordination and integration" had a substantial base well before the mass industry of the Regional Primate Research Center.

89. Carpenter field notebooks, "Asiatic Primate Expedition, Field Observations in Siam, 1937," CRC Papers. For observations on these family names, see William Domhoff, *The Higher Circles: The Governing Class in America* (New York: Random House, 1970).

90. *Comparative Psychology Monographs* 16, no. 5 (1940), reprinted in *NB*, pp. 144–271; all quotes from *NB*.
91. *NB*, p. 160.
92. CRPS Papers: 1925–23. Grantee: C. Wissler. Beatrice Blackwood was a protégée of Bronislaw Malinowski. She originally looked for a group "uncontaminated by civilization." Beatrice Blackwood, *Both Sides of Burka Passage* (London: Oxford University Press, 1935); and S[ophie] B. D. Aberle, "Frequency of Pregnancies and Birth Interval among Pueblo Indians," *American Journal of Physical Anthropology* 16 (1931): 63–80. See also Margaret Mead, *Kinship in the Admiralty Islands (New York: American Museum of Natural History, 1934).*
93. CRPS Papers: 1925–33; Grantee: C. Wissler, "Report on Anthropological Studies Supported by the CRPS, pp. 5–6.
94. Wissler, "Report," p. 6.
95. Wissler, "Report," p. 7.
96. Wissler to Angell, April 1, 1936, RMY Papers, folder 664: "Yale Laboratories of Primate Biology: Advisory Board, Wissler, C." See also A. R. Favazza and Mary Oman, *Anthropological and Cross-Cultural Themes in Mental Health, 1925–1974* (Columbia: University of Missouri Press, 1977); and Cravens, *Triumph of Evolution,* chaps. 3–4. Cravens emphasizes the goal of social control in the new human sciences, pp. 269–74. As the title of his book stresses, the product of the arrangement of the biological and social sciences since the end of the "Spencerian" chapter of the heredity-environment controversy has been an *evolutionary* human science.
97. *NB*, p. 161.
98. L. J. Henderson, *Pareto's General Sociology: A Physiologist's Interpretation* (Cambridge: Harvard University Press, 1935); and Talcott Parsons, "On Building Social System Theory: A Personal History," *Daedalus* 99 (Fall 1970): 826–81.
99. *NB*, p. 179. Carpenter cited Sherrington on neural-muscular patterns.
100. *NB*, pp. 222–23. Carpenter always approached the socionomic sex ratio from the point of view of the male; unremarkably, the bias was simply never examined.
101. J. L. Moreno, "Two Sociometries, Human and Subhuman," *Sociometry* 8 (1945): 64–75, p. 71. The journal was founded in 1937. Moreno also ran an Institute of Sociometry from New York City. J. L. Moreno, *Foundation of the Sociometric Institute* (New York: Beacon, 1942). He did his first sociograms while still in Berlin, in 1923. The range of interests can be traced in J. L. Moreno et al., eds., *The Sociometry Reader* (Glencoe, Ill.: Free Press, 1960). Sociometry was a prominent approach to the study of democratic and authoritarian personalities in the late thirties: R. Lippitt, "Field Theory and Experiment in Social Psychology: Autocratic and Democratic Group Atmosphere," *American Journal of Sociology* 45, no. 1 1939. Sociometrists were not overly modest; see J. L. Moreno, *Who Shall Survive? A New Approach to the Problem of Human Interrelations,* Nervous and Mental Disease Monograph Series no. 58 (New York: Beacon, 1934).
102. Moreno, "Two Sociometries," p. 72.
103. Moreno, "Two Sociometries," p. 72.
104. Moreno, "Two Sociometries," p. 72.
105. J. L. Moreno, Preface to Moreno et al., *Sociometry Reader,* p. vi.
106. Moreno et al., *Sociometry Reader,* p. 68.
107. M. F. Ashley Montagu, "Sociometric Methods in Anthropology," *Sociometry* 8 (1945): 62–63, p. 62 Montagu's example of successful application of sociometry was Gregory Bateson and Margaret Mead's *Balinese Character* (New York: New York Academy of Sciences, 1942). Mead, a student of Boas, was a leader in the investigation of sex, personality, and culture. Gregory Bateson, *Steps Toward an Ecology of Mind* (New York: Ballentine, 1972), was an important figure in the development of cybernetic functionalism based on analysis of communication.
108. C. R. Carpenter, "Concepts and Problems of Primate Sociometry," *Sociometry* 8 (1945): 56–61, p. 59.
109. Examples of "external" ends would be maximization of breeding potential and standardization of behavior for experimental purposes. Both these ends were discussed at length in meetings to establish primate colonies, from Yerkes's laboratories to the Regional Primate Research Centers.
110. Charles Morris, *Foundations of the Theory of Signs* (Chicago: University of Chicago Press, 1938), p. 2, my emphasis.

111. Morris, *Foundations,* p. 4.
112. Morris, *Foundations,* p. 12.
113. Morris, *Foundations,* pp. 30–32.
114. Charles Morris, *Signs, Language and Behavior* (New York: Braziller, 1955).
115. Thomas Sebeok, "Discussion of Communication Processes," in Stuart Altmann, ed., *Social Communication among Primates* (Chicago: University of Chicago Press, 1967), pp. 363–69, p. 363. For a discussion of E. O. Wilson's definition of sociobiology as a science of cybernetic communication systems, see Haraway, "Biological Enterprise," and "High Cost of Information."
116. Carpenter classified gestures in terms of form and function (e.g., affinitive, antagonistic) and immediate or long-term social control. Classifying vocalizations required considerable further technical development of sound technology. *NB,* pp. 244–55.
117. *NB,* p. 243.
118. Besides the gibbon study during A.P.E., Carpenter did a brief survey of the orang-utan in Sumatra, when Coolidge became too ill to do the work. The theme was conservation. "Animals which are preserved, frequently at large expense to governments, should be thought of as natural resources to be preserved for future generations and may be used intelligently for observational, recreational and scientific purposes." C. R. Carpenter, "A Survey of Wild Life Conditions in Atjeh, North Sumatra, with Special Reference to the Orang-utan," *Communications,* no. 12 (Amsterdam: Netherlands Committee for International Nature Protection, 1938).
119. Carpenter, "Macaques on Santiago Island, Puerto Rico"; Carpenter-Yerkes correspondence for 1937–38, RMY Papers, folder 84; Carpenter field notebooks, "The Markle-Columbia Primate Expedition," CRC Papers. C. R. Carpenter, "Rhesus Monkeys for American Laboratories," *Science* 91 (1946): 284–86.
120. Carpenter to Yerkes, October 7, 1937, RMY Papers, folder 84. In fact, standardization of laboratory primates has not been accomplished.
121. Carpenter field notebooks, "Markle-Columbia Primate Expedition," pp. 79–85.
122. Carpenter field notebooks, "Markle-Columbia Primate Expedition," p. 5.
123. Carpenter field notebooks, "Markle-Columbia Primate Expedition," p. 5.
124. Carpenter field notebooks, "Markle-Columbia Primate Expedition," p. 6.
125. Carpenter field notebooks, "Markle-Columbia Primate Expedition," p. 6.
126. Carpenter field notebooks, "Markle-Columbia Primate Expedition," p. 6.
127. Carpenter field notebooks, "Markle-Columbia Primate Expedition," p. 7.
128. Carpenter field notebooks, "Markle-Columbia Primate Expedition," p. 8.
129. Carpenter field notes, "Santiago Island," June–July 1939, February–May 1940, and May–June 1940, CRC Papers. Papers published from Cayo Santiago work were: "Sexual Behavior of Free Ranging Rhesus Monkeys, I: Specimens, Procedures, and Behavioral Characteristics of Estrus," and "II: Periodicity of Estrus, Homosexual, Anti-erotic and Non-Conformist Behavior," in *NB*, pp. 289–317 and 319–41.
130. Carpenter, "Notes on Santiago Island Colony," May–June 1940, p. 97, CRC Papers. Carpenter frequently capitalized male symbols and names. He designated females usually by a number and a lowercase symbol.
131. Carpenter, "Santiago Island Colony," p. 102.
132. Carpenter, "Santiago Island Colony," pp. 120 and 122.
133. Carpenter, "Santiago Island Colony," p. 139.
134. Carpenter, "Santiago Island Colony," p. 159.
135. Robert Redfield, ed., *Levels of Integration in Biological and Social Systems* (Lancaster, Pa.: Jacques Cattell Press, 1942); chapters included C. R. Carpenter, "Societies of Monkeys and Apes"; A. E. Emerson, "Basic Comparisons of Human and Insect Societies"; W. C. Allee, "Social Dominance and Subordination among Vertebrates"; A. L. Kroeber, "The Societies of Primitive Man"; and Robert Park, "Modern Society." "Societies of Monkeys and Apes" repeated the themes this paper has explored. Carpenter's central propositions were: "Processes of social integration and control strictly regulate the freedom of movement and limit the social contacts in organized groups of primates so as to produce definite group structure" (p. 344). "The male dominance gradient is predominant over the female gradient and acts as a kind of organizer, 'catalytic agent,' or axis for the entire group. . . . [The group is] structured with reference to dominance status of its males and to a lesser degree with reference to dominance statuses of its females" (p. 352). The

categories of group organization (central grouping tendency, socionomic sex ratio); spatial relations within and between groups; dominance; reproductive behavior and group integration; and interdependence and coordination structured the paper.

136. Redfield, *Levels of Integration,* p. 1.

137. Carpenter to Yerkes, March 20, 1939; Yerkes to Smith, March 23, 1939; Smith to Yerkes, March 27, 1939; Smith to Carpenter, March 27, 1939; and Yerkes to Carpenter, May 9, 1939, RMY Papers, folder 84. Smith also told Yerkes he and Engle did not intend to recommend Carpenter's research proposal on maternal drives and endocrines to the Committee for Research in Problems of Sex. Smith to Yerkes, March 27, 1939.

138. Titles include *Land and Life in the Jungle* and *Characterisics of the Tropics.*

139. "Audio Visual Aids Committee, Annual Report to the American Psychological Association," 1944, RMY Papers.

140. C. R. Carpenter, "Problems of German Reconstruction," *Social Action* 5 (February 1947): 29.

141. This summary of Chisholm's views and importance follows Lasch, *Haven in a Heartless World,* chap. 5, "Doctors to a Sick Society."

142. Carpenter request for a grant-in-aid from the Viking Fund, RMY Papers, folder 84.

143. Yerkes to Paul Fejos, December 1, 1947, RMY Papers, folder 84. Fejos directed the Wenner-Gren Foundation for Anthropological Research, which has been centrally instrumental in establishing primate studies in anthropology, particularly in relation to human physical, technical, and social evolution. The Wenner-Gren Foundation's Viking Fund sponsored publication of the central documents that integrated physical anthropology and primatology into the synthetic theory of evolution. Fejos was a close personal friend of Sherwood Washburn. See S. L. Washburn, ed., *Social Life of Early Man,* Viking Fund Publications in Anthropology, no. 31 (Chicago: Aldine, 1961), and *Classification and Human Evolution,* Viking Fund Publications in Anthropology, no. 37 (New York: Wenner-Gren Foundation, 1963); and F. C. Howell and F. Bourliere, eds., *African Ecology and Human Evolution,* Viking Fund Publications in Anthropology, no. 33 (New York: Wenner-Gren Foundation, 1963).

144. Information for the paragraphs to follow may be found in: Edward Yoxen, "The Social Impact of Molecular Biology," Ph.D. dissertation, Cambridge University, 1977, under the direction of Robert Young, esp. secs. 3 and 4; Claude Shannon and Warren Weaver, *The Mathematical Theory of Communication* (Urbana: University of Illinois Press, 1949) esp. Weaver, "Recent Contributions to the Mathematical Theory of Communication"; and Norbert Wiener, *Cybernetics, or Control and Communication in the Animal and the Machine* (New York: Wiley, 1948), esp. Introduction.

145. Warren Weaver, "Science and Complexity," *American Scientist* 36 (1948): 536–44.

146. Robert Kohler, "The Management of Science: The Experience of Warren Weaver and the Rockefeller Foundation Programme in Molecular Biology," *Minerva* 14 (1976): 279–306. On the material consequences for physiological functionalism in primate studies as a result of Weaver's priorities and perception of Yerkes as old-fashioned, see in RMY Papers: "Angell, 1930–49," folder 19: "Report of the Committee on Appraisal and Recommendation of the Yale Laboratories of Primate Biology, 1935," folder 665; "Annual Reports of the Anthropoid Experiment Station of Yale University," and "Administration Committee, Minutes of Meetings, 1939–42," folder 666. Weaver was also important at the Arden House meetings in 1955 in framing the terms of the American Association for the Advancement of Science's post-World War II policy on the relation of science to society. Bush played a critical role in forming the National Science Foundation. Transfer of power from the men of the generation of physiological functionalism to those who implemented cybernetic functionalism was regular. See also Steve Heims, *John von Neuman and Norbert Wiener: From Mathematics to the Technologies of Life and Death* (Cambridge: M.I.T. Press, 1980); and Haraway, "High Cost of Information."

147. A. Rosenblueth, N. Wiener, and J. Bigelow, "Behavior, Purpose and Teleology," *Philosophy of Science* 10 (1943): 18–24.

148. Wiener, *Cybernetics,* p. 17.

149. Dissemination of cybernetic approaches was well funded and systematic. Conferences bringing together linguists, logicians, physicists, physiologists, sociologists, anthropologists, and psychologists were sponsored by the Macy Foundation from 1946. Heinz von Foerster, ed., *Cybernetics: Circular Causal and Feedback Mechanisms in Biological*

*and Social Systems,* nos. 6–10 (New York: Josiah Macy, Jr., Foundation, 1950–50). Josiah Macy, Jr., Foundation, *Twentieth Anniversary Review* (New York: Macy Foundation, 1950). Rosenblueth and Wiener continued to collaborate after the war. Margaret Mead and Gregory Bateson were especially interested in the new theories for social science. Bateson, *Steps toward an Ecology of Mind.* Sociobiology would be impossible without cybernetic functionalism. One level where this dependence appears is in the analysis of information and efficiency in work distribution and division of labor in animal sociology. E. O. Wilson cites K. F. H. Murrell, *Ergonomics: Man in His Working Environment* (London: Chapman & Hall, 1965) as the source for his analysis of caste in social insects. E. O. Wilson, "The Ergonomics of Caste in Social Insects," *American Naturalist* 102 (1968): 41–66, and *Insect Societies* (Cambridge: Harvard University Press, 1971), esp. p. 342. Murrell held the meeting in his office in the British Admiralty in 1949 that laid the foundation for the interdisciplinary Ergonomics Research Society, including anatomy, physiology, anthropometry, physiological psychology, industrial medicine, physics, and engineering. Ergonomics, the word coined by Murrell's group, meant design of human-machine systems as a whole. Its first and last applications were to military problems. The 1965 book and its 1971 reprinting were attempts to popularize the point of view in industry. See also: Andras Angyal, *Foundations of a Science of Personality* (Cambridge: Harvard University Press, 1941); W. E. Duckworth, *A Guide to Operations Research* (London: Methuen, 1965); F. J. Duhl, W. Gray, and N. D. Rizzo, eds., *General Systems Theory and Psychiatry* (Boston: Little Brown, 1969); J. S. Gray, "A Physiologist Looks at Engineering," *Science* 140 (1963): 464; R. W. Jones, "An Engineer Looks at Physiology," *Science* 140 (1963): 461; Steve Heims, "Encounter of Behavioral Sciences with New Machine-Organism Analogies in the 1940s," *Journal of the History of the Behavioral Sciences* 11 (1975): 368–73; Jacob, *Logic of Life;* J. Y. Lettvin, H. R. Maturama, W. S. McCulloch, H. W. Pitts, "What the Frog's Eye Tells the Frog's Brain," *Proceedings of the Institute of Radio Engineers* 47 (1959): 1940–51 (McCulloch, Lettvin, and Pitts were very early converts to cybernetics; see Wiener, *Cybernetics,* pp. 21–23; Pitts had been a student of Carnap at Chicago); Robert Lilienfeld, *The Rise of Systems Theory* (New York: Wiley, 1978); Mesarovic, ed., *Systems Theory and Biology;* E. P. Odum, "Emergence of Ecology," and *Fundamentals of Ecology* (Philadelphia: Saunders, 1955, 1959, 1971); Michael Farley, "Formations et transformations"; Stanford L. Optner, ed., *Systems Analysis* (Baltimore: Penguin, 1973); Emanuel Peterfreund and J. T. Schwartz, *Information, Systems, and Psychoanalysis—An Evolutionary Biological Approach to Psychoanalytic Theory* (New York: International Universities Press, 1971); Henry Quastler, *Essays on the Use of Information Theory in Biology* (Urbana: University of Illinois Press, 1953); and Jagjit Singh, *Great Ideas in Information Theory, Language, and Cybernetics* (New York: Dover, 1966).

150. Yoxen, "Social Impact of Molecular Biology," chap. 9; and Norbert Wiener, *Cybernetics and Society: Human Use of Human Beings* (Boston: Houghton-Mifflin, 1950).

151. Jacob, *Logic of Life,* pp. 253–54.

152. S[tuart] A. Altmann, "Social Behavior of Anthropoid Primates: Analysis of Recent Concepts," in Eugene Bliss, ed., *Roots of Behavior* (New York: Harper, 1962). Bliss was a psychiatrist. Altmann also arranged for the translation into English of the important Japanese papers on *Macaca fuscata.* The international aspects of primatology cannot be explored here.

153. J. P. Scott, in the November 1956 *Newsletter* of the Section of Animal Behavior and Sociobiology of the Ecological Society of America, related the history of a program to stimulate animal behavior research from the founding of the Committee for the Study of Animal Societies under Natural Conditions in 1947. The Sociobiology section of the E.S.A. was formally founded in 1956.

154. A systems approach to population has been a major theme in postwar biology. Hutchinson, *Introduction to Population Ecology,* esp. chap. 3. The Ford Foundation's interest in human population control produced some funding for study of wild primate populations. Ernest Knobil, Chairman of the Department of Physiology, University of Pittsburgh School of Medicine, wrote to Carpenter after an exploratory trip to southeast Asia: "The Ford Foundation, under the aegis of its Population Program, has undertaken to support several projects devoted to various aspects of primate reproduction. In this connection, it has also been interested in the possibility of establishing a large primate laboratory devoted to reproductive physiology somewhere in Asia." Carpenter responded:

"I, too, am much interested in the reproduction and population aspects of primate behavior." Knobil to Carpenter, November 29, 1965; and Carpenter to Knobil, December 9, 1965, CRC Papers, "Primate Centers, Puerto Rico" folders.

155. David McK. Rioch, Discussion for pt. 2, "Agonistic Behavior," in Altmann, ed., *Social Communication among Primates.*

156. Altmann to Carpenter, February 5, 1956, CRC papers, "Monkey and Other Animal Studies" folder. Rioch had consulted with Robert Morison, Director of Biological and Medical Research at the Rockefeller Foundation, on the fate of Cayo Santiago. Morison to Carpenter, November 23, 1955, CRC Papers, "Monkey and Other Animal Studies" folder. Rioch had sent a letter on January 6, 1956 to leaders in American animal behavioral studies on the problems of Cayo Santiago: "It seems to me that the main problem is to find an adequate man [sic] to head up the project. If one is not available in the United States, possibly we should try to get a student from Lorenz or from Tinbergen." Rioch to Carpenter, January 6, 1956, CRC Papers, "Primate Committee, Institute of Animal Resources" folder. In 1956 Cayo Santiago came under the direction of the National Institutes of Neurological Diseases and Blindness. Management of Cayo Santiago has been complicated and is not critical to the theme of this paper. Agreement between the National Institute of Neurological Diseases and Blindness and the University of Puerto Rico, April 16, 1956. CRC Papers, "Primate Centers, Puerto Rico" folder.

157. Carpenter to Altmann, June 5, 1956, CRC Papers, "Monkey and Other Animal Studies" folder.

158. Enclosure from Rioch to Carpenter and others, January 6, 1956, CRC Papers, "Primates Committee, National Academy of Sciences, Institute of Animal Resources" folder.

159. Haraway, "Past Is the Contested Zone," esp. pp. 50–53. While chairman of the Department of Psychiatry of Stanford University's Medical School, Hamburg arranged the collaboration of Jane Van Lawick-Goodall in studying chimpanzees in Gombe East and Gombe West—Zaire and Palo Alto. Goodall's work was permeated with psychiatric and personality adjustment concerns. Carpenter also would have liked to induce Goodall to associate with the Yerkes Primate Research Center in Atlanta, but he could not command the research funds, especially from the National Institutes of Mental Health, that Hamburg could. Carpenter to Van Lawick-Goodall, May 12, 1971; Carpenter to Teleki, January 3, 1969; and Carpenter to Van Lawick-Goodall, September 4, 1970, CRC Papers, "Geza Teleki" folder and "Primates—General Correspondence" folder. Goodall's doctoral degree was arranged under the direction of Robert Hinde, of Cambridge University, who remained as scientific adviser to the Gombe field station. Hinde was a major figure in the synthesis of American and British comparative psychology and continental ethology and has done important work on primate communications, socialization, and social role. Robert Hinde, *Animal Behavior* (New York: McGraw-Hill, 1966), *Nonverbal Communication* (London: Cambridge University Press, 1972), *Biological Bases of Human Social Behavior* (New York: McGraw-Hill, 1974), and "The Relevance of Animal Studies to Human Neurotic Disorders," in Richter et al., eds., *Aspects of Psychiatric Research* (London: Oxford University Press, 1962), pp. 240–61.

160. Altmann, rough draft of research, CRC papers, March 22, 1956, "Monkey and Other Animal Studies" folder. Altman was E. O. Wilson's first graduate student.

161. E. O. Wilson, *Sociobiology* (Cambridge: Harvard University Press), 1975.

162. Altmann, ed., *Social Communication among Primates;* and DeVore, ed., *Primate Behavior.*

163. Sebeok, "Discussion of Communication Processes," in Altmann, ed., *Social Communication among Primates,* p. 367.

164. See also C. F. Hockett and R. Ascher, "The Human Revolution," *Current Anthropology* 5 (1964): 135–47; C. F. Hockett, *A Course in Modern Linguistics* (New York: Macmillan, 1958); Peter Marler, "The Logical Analysis of Animal Communication," *Journal of Theoretical Biology* 1 (1961): 295–317; and T. A. Sebeok, A. S. Hayes, and M. C. Bateson, *Approaches to Semiotics* (The Hague: Mouton, 1964).

165. Stuart A. Altmann, ed., *Japanese Monkeys: A Collection of Translations* (Edmonton, Ill., 1965).

166. The Primate Research Group of Kyoto University, "Field Notes on the Social Behavior of Japanese Monkeys *(Macaca fuscata),*" Kyoto, Japan, 1956, CRC Papers,

Monkey and Other Animal Studies folder. Among the critical figures in the group were K. Imanishi, S. Kawamura, and J. Itani. From the beginning of their observations of semiwild horses (Imanishi) and deer (Kawamura), the Japanese adopted the method of rigorous identification of individual animals and intensive study of their social interactions. Their longitudinal studies of monkey groups composed entirely of known individuals are the most extensive and long-standing in the world. The Washburn lineage in primate studies was also interested early in the Japanese investigation, and a member of Washburn's seminar at the University of Chicago, John Frisch, published an important introduction to their work: "Research on Primate Behavior in Japan," *American Anthropologist* 61 (1959): 584–96. The seminar was part of the Ford Foundation–sponsored project on Evolution of Human Behavior, directed by Washburn.

167. Carpenter, "Notebook, Japan Trip, Aug. 22–Oct. 1, 1966," CRC Papers.

168. Conference at the University of Missouri, May 2–3, 1969, CRC Papers, "Centers: Primate Centers, Japanese Macaques" folder. Peter Marler is another important figure in development of sociobiology. Peter Marler, "Animal Communication Signals," *Science* 157 (1967): 769–74; and Peter R. Marler and W. J. Hamilton, *Mechanisms of Animal Behavior* (New York: Wiley, 1966). Like many of the primatologists, Marler was interested in popularization of the story of animal communication. Peter Marler, ed., *The Marvels of Animal Bahavior* (Washington, D.C.: National Geographic Society, 1972).

169. G. B. Schaller, *The Mountain Gorilla: Ecology and Behavior* (Chicago: University of Chicago Press, 1963).

170. CRC Papers, African Primate Studies folder, esp. correspondence among Emlen, Carpenter, Southwick, Coolidge, Schaller, and Washburn. Emlen was instrumental in getting both Charles Southwick and William Mason interested in a broad primate program, including population dynamics. Emlen to Carpenter, November 17, 1958. Southwick and Mason both participated with Carpenter in the 1959 census of howlers considered further below. Mason is a link with the Harlow work at the University of Wisconsin; Southwick, formerly located at the School of Hygiene and Public health at Johns Hopkins and now at the University of Colorado, is important in study of population dynamics and cropping potential of primates, especially rhesus in India.

171. Gordon Stephenson and John Emlen, "Group-Specific Communication Patterns in Primates: II," grant proposal for period of September 1, 1971 through August 31, 1974, pp. 1–2, CRC Papers.

172. S[tuart] A. Altmann, "Sociobiology of Rhesus Monkeys. II: Stochastics of Social Communication," *Journal of Theoretical Biology* 8 (1965): 490–522, and "IV: The Basic Communication Network," *Behaviour* 32 (1968): 17–32.

173. Stephenson and Emlen, "Group-Specific Communication Patterns," p. 17.

174. Stephenson and Emlen, "Group-Specific Communication Patterns," p. 15.

175. Stephenson and Emlen, "Group-Specific Communication Patterns," p. 16.

176. Stephenson and Emlen, "Group-Specific Communication Patterns," p. 39. The Emlen-Stephenson proposal noted arrangements for sophisticated analysis of 196,000 feet of film. Marler, at the Rockfeller University, helped here.

177. Marshall Sahlins, *The Use and Abuse of Biology* (Ann Arbor: University of Michigan Press, 1976).

178. Wilson, *Sociobiology,* chap. 2.

179. Cravens, *Triumph of Evolution,* chap. 3.

180. Cravens, *Triumph of Evolution,* pp. 269ff.

181. Ernst Mayr, "The Evolution of Living Systems," *Proceedings of the National Academy of Sciences* 51 (1964): 934–41. Theodosius Dobzhansky, *The Biological Basis of Human Freedom* (New York and London: Columbia University Press, 1956).

182. Stephenson and Emlen, "Group-Specific Communication Patterns," pp. 8 and 11.

183. Altmann held the position of "Sociobiologist" at the Yerkes Regional Primate Research Center from 1965 to 1970, when he became professor of biology and anatomy at the University of Chicago. The crisscrossing of personnel and institutions has been a leitmotif in the primate story.

184. Carpenter, "Social Behavior of Nonhuman Primates," in *NB,* p. 371; and C. R. Carpenter, "Field Studies of a Primate Population," in Bliss, ed., *Roots of Behavior,* and *NB,* p. 400.

185. Carpenter, "Social Behavior of Nonhuman Primates," in *NB,* p. 377, and "Field Studies of a Primate Population," in *NB,* p. 399.

186. Carpenter, "Social Behavior of Nonhuman Primates," in *NB,* p. 380.

187. Carpenter, "Social Behavior of Nonhuman Primates," in *NB,* p. 381.

188. Stuart A. Altmann, "The Structure of Primate Social Communication," in Altmann, *Social Communication among Primates,* pp. 325–62.

189. Carpenter to Ronald Meyers, December 14, 1965; Carpenter to Hockett, January 6, 1966; Morrison to Geoffrey Bourne, of the Yerkes Primate Research Center, March 14, 1966; Bourne to Morrison, March 23, 1966; Carpenter to Morrison, April 12, 1966; Morrison to Carpenter, April 26, 1966; and Carpenter to Morrison, May 9, 1966, CRC Papers, "Primate Centers, Puerto Rico" folder. Meyers was director of the laboratory in 1966; John Morrison was acting director of the Laboratory of Perinatal Physiology, which was responsible for the monkey colony.

190. Carpenter, "Field Studies of a Primate Population," in *NB,* pp. 400–405.

191. Carpenter, "Field Studies of a Primate Population," in *NB,* p. 406. The popular writer Robert Ardrey extensively used Carpenter's work on territory, aggression, and population equilibrium. Ardrey and Carpenter carried on a long and mutually enthusiastic correspondence. The relation of popularizer and expert was important to the ideological function of the primate story in promoting the proper sorts of healthy adjustment. Ardrey referred a New York labor arbitration official to Carpenter for expert advice in relation to applying the concept of territoriality to settling labor disputes. Carpenter to Donald Strauss, of the American Arbitration Association, November 2, 1966; Strauss to Ardrey, September 29, 1966; and Strauss to Carpenter, October 25, 1966, CRC Papers, "Robert Ardrey file, 1964–70" folder. Carpenter referred worried white parents during the southern civil rights struggle to Ardrey. Dorothy Mayhue to Carpenter, March 16, 1970; and Carpenter to Mayhue, April 20, 1970, CRC Papers, "Primates, General Correspondence" file. See esp. Robert Ardrey, *The Social Contract: A Personal Inquiry into the Evolutionary Sources of Order and Disorder* (New York: Atheneum, 1970). Carpenter worked hard to promote this book; he tried to get his favorable review published in several popular magazines. *Life* had serialized the book, hoping that its controversial content would help the financially troubled magazine. *Life*'s special interest was population control: Ardrey to Carpenter, March 21, 1970. The BBC was enthusiastic about making thirteen 50-minute features written by Ardrey on a theme such as the origin of violence. Ardrey to Carpenter, July 23, 1970; Carpenter to Ardrey, October 23, 1970; and Carpenter to Ardrey, September 9, 1970. At the time Ardrey finished *Territorial Imperative* in 1966, he wrote the film script for *Khartoum,* a $7 million United Artists extravaganza. *Territorial Imperative* was especially popular in secondary schools. Ardrey to Carpenter, January 27, 1967. Carpenter's approval of Ardrey continued only as long as the latter maintained his position as translator of science and did not usurp the superior status of his scientist-sources. "Finally I must report that I had a two hour conversation with Robert Ardrey in Atlanta recently and found him the same *egocentric authority* that he was previously. I think he is making a mistake by trying to become the *authority* rather than the *scribe* and interpreter." Carpenter to Alan Elms, Psychology Department, University of California at Davis, January 11, 1973. All correspondence in CRC Papers, "Robert Ardrey file, 1964–70" folder.

192. Hans Selye, *The Stress of Life* (New York: McGraw-Hill, 1956), tells the story of the engineering of the stress concept. Lawrence E. Hinkle, "The Concept of 'Stress' in the Biological and Social Sciences," *Science, Medicine, and Man* 1 (1973): 31–48, discusses the transformation of stress into a cybernetic concept central to medicine and psychiatry.

193. C. R. Carpenter, "Territoriality, A Review of Concepts and Problems," in *NB,* p. 421. See also Roe and Simpson, *Behavior and Evolution.*

194. C. R. Carpenter, "Aggressive Behavior of Organisms, Including Man: The Need for an Integral System of Information," CRC Papers. The printed version appeared in M. Fried, M. Harris, and R. Murphy, eds., *War: The Anthropology of Armed Conflicts and Aggression* (Garden City, N.Y.: Natural History Press, 1968).

195. Carpenter, "Aggressive Behavior of Organisms," p. 7.

196. Carpenter, "Aggressive Behavior of Organisms," p. 12.

197. Carpenter, "Aggressive Behavior of Organisms," p. 12.

198. C. R. Carpenter, ed., *Behavioral Regulators of Behavior in Primates* (East Brunswick, N.J.: Associated University Presses, 1973), p. 9.

199. Carpenter, ed., *Behavioral Regulators,* p. 16. Primatology was one of the areas sponsored by the Japan–United States Cooperative Science program, in which Carpenter participated.

200. C. R. Carpenter, "The Application of Less Complex Instructional Technologies," in Wilbur Schramm, ed., *Quality Instructional Television* (Honolulu: Hawaii University Press, East-West Center, 1972), p. 191.

201. Carpenter, "Less Complex Instructional Technologies," p. 192. The language is Carpenter's.

202. Carpenter, "Less Complex Instructional Technologies," p. 200.

203. Carpenter, "Less Complex Instructional Technologies," p. 201.

204. Carpenter, "Less Complex Instructional Technologies," p. 203, emphasis in original.

205. Carpenter, "Less Complex Instructional Technologies," p. 204. For a "total systems approach" to educational television in the United States, for which Carpenter worked from the first national conference sponsored by the Joint Council on Educational TV in 1952 at Pennsylvania State, see John W. Meaney and C. R. Carpenter, eds., *Telecomnunications: Toward National Policies for Education* (Washington, D.C.: Joint Council on Educational Telecommunications, 1970).

206. "Telemetric Control of Free-Ranging Gibbons," Public Health Service Grant Application, Aristide Esser, Principal Investigator, CRC Papers.

207. C. R. Carpenter, "Suggestions for the Development of a Gibbon Colony for Behavioral and Neurophysiological Research," CRC Papers.

208. "Telemetric Control of Free-Ranging Gibbons," p. 18.

209. "Telemetric Control of Free-Ranging Gibbons," p. 19. Controversy over choice of the right ape to model the preferred human aspect provides moments of low comedy for the historian. Those who wish to stress a male role in enforcing social control and in evolving "cooperation" on the dangerous savannah after expulsion by drought from the tropical garden generally choose baboons. Those who want to highlight matrifocal groups and sociobiological explanations of a peaceful human nature favor chimpanzees. Nancy Tanner and Adrienne Zihlman, "Women in Evolution," *Signs* 1, no. 3, pt. 1 (1976): 585–608. This author enjoys the last strategy. Washburn has commented wryly on this deep current in his discipline in his article "Human Behavior." It is a serious fact that twentieth century primatologists make constant use of biblical mythology in the structure of their primate sources. These stories permeate work and ideas in both popular and technical contexts. See for example, Robin Fox, "In the Beginning: Aspects of Hominid Behavioral Evolution," *Man* 2 (1967): 415–33.

210. "Telemetric Control of Free-Ranging Gibbons," p. 27.

211. "Telemetric Control of Free-Ranging Gibbons," pp. 22, 23, and 27. This research was part of the background for Marge Piercy's novel, *Woman on the Edge of Time* (New York: Knopf, 1976), reviewed by Donna Haraway, "The Search for a Feminist Science," *Women: A Journal of Liberation* 6, no. 2 (1979): 20–23. Relevant publications by Delgado and Esser are: A. H. Esser, "Interactional Hierarchy and Power Structure on a Psychiatric Ward: Ethological Studies on Dominance Behavior in a Total Institution," in C. Hutt and S. J. Hutt, eds., *Behavioral Studies in Psychiatry* (Oxford: Pergamon Press, 1970); A. H. Esser, A. S. Chamberlain, E. D. Chapple and N. S. Kline, "Territoriality of Patients on a Research Ward," in J. Wortis, ed., *Recent Advances in Biological Psychiatry* 7 (New York: Plenum, 1965): J. M. R. Delgado, "Personality, Education and Electrical Stimulation of the Brain," in A. E. Traxler, ed., *Innovation and Experiment in Modern Education* (Washington: American Council on Education, 1965), pp. 121–29; J. M. R. Delgado, "Brain Technology and Psychocivilization," in C. P. Hall, ed., *Human Values and Advancing Technology,* A New Agenda for Church in Mission (New York: Friendship Press, 1967), pp. 68–92; and J. M. R. Delgado, V. Mark, W. Sweet, F. Ervin, G. Weiss, G. Bach-y-Rita, and R. Hagiwara, "Intercerebral Radio Stimulation and Recording in Completely Free Patients," *Journal of Nervous and Mental Disorders* 147 (1968): 329–40. The connection of psychiatry and ethology, favored by Delgado and Esser, has been especially pursued in England.

212. Carpenter to Margaret Varley, May 3, 1968, CRC Papers, "Primate Centers, Puerto Rico" folder.

213. Varley to Carpenter, May 17, 1968 and August 6, 1968; Carpenter to Varley, May 24, 1968 and August 27, 1968; E. A. Missakian to Carpenter, May 21, 1968 and June 18, 1968; and Carpenter to Missakian, June 7, 1968, CRC Papers.

214. Carpenter Bermuda notebook, and "Suggestions for the Gibbon Colony," CRC Papers.

215. Yerkes, *Chimpanzees: A Laboratory Colony,* chap. 1, "Servant of Science." Carpenter died in 1974, in Athens, Georgia.
216. Useful places to witness questioning of the universality and relevance of the nature/culture distinction are Carol MacCormack and Marilyn Strathern, eds., *Nature, Culture, and Gender* (London and New York: Cambridge University Press, 1980); and Edward Said, *Orientalism* (N.Y.: Pantheon, 1978).
217. Wiseman, *Primate,* 1974, distributed by Xippora Films, Boston. Wiseman also made *High School, Hospital, Meat, Basic Training, Panama Canal, Juvenile Court, Welfare,* and *Titicut Follies.* His films are seen on public television; *Primate* was aired in 1974. I have not met a primatologist who likes it; they have good reason. Wiseman takes on the major institutions of modern life in his documentaries.
218. Eugene Marais, *Soul of the White Ant,* version by South Africa Radio Broadcasting, 1980. Marais was also an early observer of baboons. His organicist-superorganismic views parallel those of William Morton Wheeler and Alfred Earl Emerson.
219. Latour and Woolgar, *Laboratory Life,* explore fully the notion of an inscription device in the system of scientific fact production. See esp. chap. 2, "An Anthropologist Visits the Laboratory." They draw from Jacques Derrida, *Of Grammatology* (Baltimore: Johns Hopkins University Press, 1977).
220. For Bourne's own presentation of the work and meanings of the Yerkes Primate Research Center, see Geoffrey Bourne, *The Ape People* (New York: G. P. Putnam's Sons, 1971).